T0237704

Artificial Intelligence: Foundations, Theory, and Algorithms

Series Editors

Barry O'Sullivan, Dep. of Computer Science, University College Cork, Cork, Ireland

Michael Wooldridge, Department of Computer Science, University of Oxford, Oxford, UK

Artificial Intelligence: Foundations, Theory and Algorithms fosters the dissemination of knowledge, technologies and methodologies that advance developments in artificial intelligence (AI) and its broad applications. It brings together the latest developments in all areas of this multidisciplinary topic, ranging from theories and algorithms to various important applications. The intended readership includes research students and researchers in computer science, computer engineering, electrical engineering, data science, and related areas seeking a convenient way to track the latest findings on the foundations, methodologies, and key applications of artificial intelligence.

This series provides a publication and communication platform for all AI topics, including but not limited to:

- Knowledge representation
- Automated reasoning and inference
- Reasoning under uncertainty
- Planning, scheduling, and problem solving
- Cognition and AI
- Search
- Diagnosis
- Constraint processing
- Multi-agent systems
- Game theory in AI
- Machine learning
- Deep learning
- Reinforcement learning
- Data mining
- Natural language processing
- Computer vision
- Human interfaces
- Intelligent robotics
- Explanation generation
- Ethics in AI
- Fairness, accountability, and transparency in AI

This series includes monographs, introductory and advanced textbooks, state-of-the-art collections, and handbooks. Furthermore, it supports Open Access publication mode.

Xiaowei Huang • Gaojie Jin • Wenjie Ruan

Machine Learning Safety

 Springer

Xiaowei Huang
University of Liverpool
Liverpool, UK

Gaojie Jin
University of Liverpool
Liverpool, UK

Wenjie Ruan
University of Exeter
Exeter, UK

ISSN 2365-3051 ISSN 2365-306X (electronic)
Artificial Intelligence: Foundations, Theory, and Algorithms
ISBN 978-981-19-6816-7 ISBN 978-981-19-6814-3 (eBook)
https://doi.org/10.1007/978-981-19-6814-3

This Springer imprint is published by the registered company Springer Nature Singapore Pte Ltd.
The registered company address is: 152 Beach Road, #21-01/04 Gateway East, Singapore 189721,
Singapore

Preface

This book addresses the safety and security perspective of machine learning, focusing on its vulnerability to environmental noise and various safety and security attacks. Machine learning has achieved human-level intelligence in long-standing tasks such as image classification, game playing, and natural language processing. However, like other complex software systems, it is not without any shortcomings, and a number of hidden issues have been identified in the past years. The vulnerability of machine learning has become a major roadblock to the deployment of machine learning in safety-critical applications.

We will first cover falsification techniques to identify the safety vulnerabilities on various machine learning models, and then devolve them into different solutions to evaluate, verify, and reduce the vulnerabilities. The falsification is mainly done through various attacks such as robustness attacks, data poisoning attacks, etc. Compared with the popularity of attacks, solutions are less mature, and we consider solutions that have been broadly discussed and recognised (such as formal verification, adversarial training, and privacy enhancement), together with several new directions (such as testing, safety assurance, and reliability assessment).

Specifically, this book includes four technical parts. Part I introduces basic concepts of machine learning, as well as the definitions of its safety and security issues. This is followed by the introduction of techniques to identify the safety and security issues in machine learning models (including both transitional machine learning models and deep learning models) in Part II. Then, we present in Part III two categories of safety solutions that can verify (i.e. determine with provable guarantees) the robustness of deep learning and that can enhance the robustness, generalisation, and privacy of deep learning. In Part IV, we discuss several extended safety solutions that consider either other machine learning models or other safety assurance techniques. We also include technical appendices.

The book aims to improve the awareness of the readers, who are future developers of machine learning models, on the potential safety and security issues of machine learning models. More importantly, it includes up-to-date content regarding the safety solutions for dealing with safety and security issues. While these solution techniques are not sufficiently mature by now, we are expecting that they can be further developed, or can inspire new ideas and solutions, towards the ultimate goal of making machine learning safe. We hope this book can pave the way for the readers to become researchers and leaders in this new area of machine learning safety, and the readers will not only learn technical knowledge but also gain hands-on practical skills. Some source codes and teaching materials are made available at https://github.com/xiaoweih/AISafetyLectureNotes.

London, UK   Xiaowei Huang
Liverpool, UK   Gaojie Jin
Exeter, UK   Wenjie Ruan
March 2022

Contents

Acronyms

AI	Artificial Intelligence
AUC	Area Under Curve
CNN	Convolutional Neural Networks
DNN	Deep Neural Networks
DRL	Deep Reinforcement Learning
LP	Linear Programming
MAP	Maximum a Posteriori
MILP	Mixed Integer Linear Programming
PR Curve	Precision/Recall Curve
ROC Curve	Receiver Operating Characteristic Curve

Part I
Safety Properties

The first part of this book will introduce fundamental knowledge about machine learning (Chap. 1) and discuss how the machine learning models are evaluated, including both traditional model evaluation methods (Chap. 2) and the focus on this book—safety properties (Chap. 3). For safety properties, we only present their definitions, with the methods on how to deal with them covered throughout the book. The readers are referred to Part 4 (Appendix) for mathematical foundations.

While this book includes content on the design and training of machine learning models, its key focus is on whether a trained machine learning model will perform safely when deployed in a real-world application. For example, it is interesting to know if a perception system, implemented with convolutional neural networks, can work well in a self-driving car system without compromising its safety through e.g., misclassifying the pedestrians or a lorry. Model evaluation (Chap. 2) is traditionally an integral part of the machine learning model development process. It uses statistical methods to help determine the best machine learning model for a given dataset, and help understand how well the machine learning model will perform in the future. All the evaluations are dependent on the dataset that is collected prior to the model development process. While model evaluation methods give some indications on the quality of a machine learning model, they are not testing whether or not the model is "correct". Rather, it considers whether the model "fits" well for the training data, or whether the model is "useful" for the problem.

Safety properties (Chap. 3) describe the safety and security errors when deploying a machine learning model in an application, in particular when the application may present some risks that are not present in (or cannot be easily detected from) the training dataset. In this book, we consider safety and security as interchangeable concepts, both of which suggest the system is free of risks, with security focusing on deliberate attacks from an attacker. The properties to be discussed will include the consideration that the training dataset is not representative enough for the actual working environment (e.g., generalisation error), the consideration that the working environment may include noises (e.g., robustness error), and the consideration that the working environment may have adversarial agents that intend to compromise the machine learning model for their benefits (e.g., adversarial examples, poisoning attacks, backdoor attacks, model stealing, membership inference, and model inversion).

After the definition of safety and security properties in this part, we will discuss the safety threads, i.e., the identification of safety and security errors, in Part II. We will then discuss two core safety solutions, i.e., verification and enhancement, in Part III, and several extended safety solutions in Part IV.

Chapter 1
Machine Learning Basics

This chapter presents basic concepts and definitions about machine learning, including data representation, dataset, hypothesis space, inductive bias, and various learning tasks and learning schemes. Moreover, we will also discuss density estimation, ground truth, and underlying data distribution.

1.1 What Is Machine Learning?

A machine learning algorithm is a software program that can improve its performance by learning from data (or examples). As shown in Fig. 1.1, given a program (or model) f, the improvement (from f to f') is achieved by applying a learning algorithm on f and a set D_{train} of training instances.

Usually, a learning algorithm cannot fully comprehend the training instances with a single pass. Therefore, the learning over a dataset D_{train} is an iterative process, i.e., the learning step shown in Fig. 1.1 is repeated until a termination condition is satisfied. A termination condition can be e.g., a number of iterations, an accuracy threshold, and a convergence condition.

Based on the above process, and depending on the nature of the data instances and how the learning algorithm interacts with the data instances, there can be different learning tasks and learning schemes. We will briefly discuss the basic categories of them in Sects. 1.5 and 1.6, leaving the details of the learning algorithms to Part II and Part 10. In the following, we explain the data representation (Sect. 1.2.1) and the datasets (Sect. 1.3).

© The Author(s), under exclusive license to Springer Nature Singapore Pte Ltd. 2023
X. Huang et al., *Machine Learning Safety*, Artificial Intelligence: Foundations,
Theory, and Algorithms, https://doi.org/10.1007/978-981-19-6814-3_1

Fig. 1.1 Machine learning

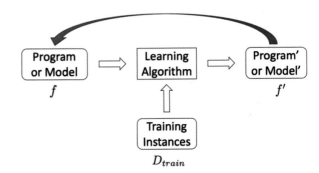

1.2 Data Representation

Data representation refers to the way how the data instances are stored and represented. A suitable data representation can ease the storage, transmission, and processing of data, and more importantly, benefit the machine learning algorithms which heavily rely on not only the data but also the way how the data is operated.

1.2.1 Representing Instances Using Feature Vectors

A dataset is formed of a finite set of data instances as well as their associated labels if available. One common way to represent a data instance is to use a fixed-length vector $\mathbf{x}$ to represent the features (or attributes) of the data instance. Standard feature types include e.g.,

- nominal (including Boolean) type, such that there is no ordering among possible values of the feature. For example, $color \in \{red, blue, green\}$ is a nominal feature.
- ordinal type, such that possible values of the feature are totally ordered. For example, $size \in \{small, medium, large\}$ is an ordinal type.
- numeric (continuous) type, whose values are stored as groupings of bits, such as bytes and words. Numbers, such as integers and real numbers, are typical examples of numeric types. As an example, $weight \in [0 \ldots 500]$ is a numeric type.

Example 1.1 For the **iris** dataset [38], we have the information in Fig. 1.2a, b. The left picture (Fig. 1.2a) is an illustration of an iris flower, where we can see the intuitive meanings of four features: sepal length, sepal width, petal length, and petal width. The right table (Fig. 1.2b) is a snapshot of the dataset, which has in

a **b**

index	Sepal Length	Sepal Width	Petal Length	Petal Width	Class Label
1	5.1	3.5	1.4	0.2	iris setosa
2	4.9	3.0	1.4	0.2	iris setosa
...					
50	6.4	3.5	4.5	1.2	iris versicolor
...					
150	5.9	3.0	5.1	1.8	iris virginica

Fig. 1.2 (**a**) An iris flower. (**b**) Iris dataset

total of 150 instances. We can see that, all four features are represented as numeric values.

We may write $\mathbf{x} = (x_1, \ldots, x_n)$ for an instance with n features, each of which has value x_i for $i \in \{1, \ldots, n\}$. Then, in case we are dealing with labelled data, we also need to represent the label of each instance $\mathbf{x}$. Depending on the nature of the problem, a label can be represented as either a scalar y (e.g., classification) or a vector $\mathbf{y}$ (e.g., object detection). There are also tasks in which each label is a structured object.

Example 1.2 For the **iris** example, we can see from Fig. 1.2b that each instance is associated with a label indicating it is in one class (3 classes in total). If we use 1 for iris setosa, 2 for iris versicoler, and 3 for iris virginica, we may have $\mathbf{x}_1 = (5.1, 3.5, 1.4, 0.2)$, and $y_1 = 1$.

Example 1.3 In the *object detection* task, a label is a set of bounding boxes, each of which is associated with not only a classification label but also other information such as the size of the box, the coordinates of the box in the image, etc.

In the following, we may also write $(\mathbf{x}, y)$ for an instance, assuming that the label is a scalar number y.

1.2.2 Feature Space

We can think of each instance $\mathbf{x}$ as representing a point in a n-dimensional feature space where n is the number of features of $\mathbf{x}$.

Example 1.4 Assume that we are working with a dataset which is sampled from the following underlying function:

$$
\begin{aligned}
X &\in [0, 7] \\
Y &= \sin 7X + \epsilon \\
Z &= \cos 7X + \epsilon \\
\epsilon &\in [0, 1]
\end{aligned}
\tag{1.1}
$$

such that all instances contain 3 numerical features X, Y, Z. ϵ denotes the noise. Each data instance is a point in a 3-dimensional space. The visualisation of the dataset is seen in Fig. 1.3. The code for the generation of the figure is given in Chap. 4.

Assume that all features are numeric and every feature X_i is in a set R_i. Then, the feature space is $\prod_{i=1}^{n} R_i = R_1 \times \ldots \times R_n$.

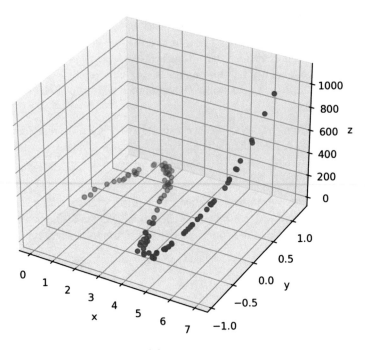

Fig. 1.3 Visualisation of a 3-dimensional dataset

1.3   Datasets

A dataset is a collection of data instances. According to their different roles in the machine learning lifecycle, we may have multiple datasets, including training dataset, test dataset, and validation dataset.

1.3.1   Independent and Identically Distributed (i.i.d.) Assumption

We often assume that training instances are independent and identically distributed (i.i.d.), i.e., the instances are sampled from the same probability distribution (i.e., identically distributed) and all of them are mutually independent. i.i.d. assumption has an important property for a valid dataset, because it enables the following property:

> the larger the size of the dataset becomes, the greater the probability of the data instances will closely resemble the underlying distribution.

We remark that, this property is key to many machine learning theoretical results based on e.g., Chebyshev's inequality, and Hoeffding's lemma.

However, there are also cases where this assumption does not hold, such as

- instances sampled from the same medical image,
- instances from time series,
- etc.

For non-i.i.d. datasets, dedicated techniques have to be taken to make sure the learning algorithm can learn useful information instead of biased information.

1.3.2   Training, Test, and Validation Datasets

The collected dataset can be split into two datasets, one for training and the other for test. We call them *training dataset* and *test dataset*, respectively. The split does not have a fixed percentage, with e.g., 7:3 or 8:2 split being very commonly seen in practice. Within the training dataset, there might be a subset called *validation dataset* that is often used to control the learning process, such as the termination of the learning process or the selection of the learning directions and hyper-parameters. Figure 1.4 presents an illustrative diagram for the three datasets.

Fig. 1.4 Training, test, and validation datasets

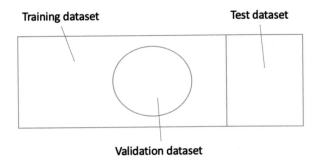

Training dataset   Test dataset

Validation dataset

1.4   Hypothesis Space and Inductive Bias

As suggested in Fig. 1.1, a learning agent f is a program or model or function. It updates itself by learning from training data. Nevertheless, the function f has to be chosen from a function space $\mathcal{H}$, called hypothesis space. Normally, the hypothesis space is determined by the learning algorithm.

Example 1.5 For a decision tree algorithm, the hypothesis space is the set of all possible trees such that each tree node represents a feature and the branches of a tree node represent the split of the possible feature values.

Example 1.6 For a linear regression algorithm, the hypothesis space is the set of linear functions $y = \mathbf{w}^T\mathbf{x} + b$, where $\mathbf{w} \in \mathbb{R}^{|\mathbf{x}|}$ and $b \in \mathbb{R}$.

Example 1.7 For a neural network whose corresponding function $f_{\mathbf{W}}$ is parameterised over learnable weights $\mathbf{W}$, the hypothesis space is the set of functions $f_{\mathbf{W}}$ such that each weight in $\mathbf{W}$ is a real number.

The inductive bias (also known as learning bias) of a learning algorithm is the set of assumptions that the designer imposes on the hypotheses in $\mathcal{H}$ to guide the learner in its learning. Usually, the assumptions can be e.g.,

- restrictive assumption that limits the hypothesis space, or
- preference assumption that imposes ordering on hypothesis space, etc.

Example 1.8 For decision tree learning, it is possible to ask for a preference over simpler trees, by following Occam's razor.

Example 1.9 For neural networks, it is possible to apply regularisation techniques, such as L_1 or L_2 regularisation, so that the learning algorithm will have a preference between weight matrices $\mathbf{W}$ embedded.

1.5 Learning Tasks

Given a dataset, we also need to determine the learning task before applying machine learning algorithms. Different machine learning algorithms (and probably datasets) will be needed, according to different learning tasks.

1.5.1 Supervised Learning

One of the most frequently seen tasks is supervised learning, which learns a function according to a set of input-output pairs. The primary objective in supervised learning is to make sure that the learned function generalises, i.e., it is able to accurately predict label y for previously unseen $\mathbf{x}$.

Definition 1.1 (Supervised Learning) Given a training set of instances sampled from an unknown target function h, i.e.,

$$D = \{(\mathbf{x}^{(1)}, y^{(1)}), (\mathbf{x}^{(2)}, y^{(2)}), \ldots, (\mathbf{x}^{(m)}, y^{(m)})\} \tag{1.2}$$

it is to learn a function $f \in \mathcal{H}$ that approximates the target function h, where $\mathcal{H}$ is a set of models (a.k.a. hypotheses). We call D a labelled dataset when each instance $\mathbf{x}$ is attached with a label y, and unlabelled, otherwise.

In the above definition, when y is discrete, it is a *classification* task (or concept learning). When y is continuous, it is a *regression* task. The function f is called classifier or regressor, depending on the tasks. For a classifier, it is often that, instead of directly returning a label, $f(\mathbf{x})$ returns a probabilistic distribution over the set C of labels such that

$$\sum_{c \in C} f(\mathbf{x})(c) = 1 \tag{1.3}$$

and we let its label be the one with maximum probability, i.e.,

$$\hat{y} = \arg\max_{c \in C} f(\mathbf{x})(c) \tag{1.4}$$

Note that, a predictive label $\hat{y}$ may be different from the ground truth label y.

1.5.2 Unsupervised Learning

Another popular learning task is unsupervised learning, which, instead of asking users to supervise the learning with example input-output pairs, asks the learning algorithm to discover patterns and information that were previously undetected. Formally,

Definition 1.2 (Unsupervised Learning) Given a set of training instances without y's, i.e., an unlabelled dataset

$$D = \{\mathbf{x}^{(1)}, \mathbf{x}^{(2)}, \ldots, \mathbf{x}^{(m)}\} \tag{1.5}$$

it is to discover interesting regularities (such as structures and patterns) that characterise the instances.

Concretely, depending on the "interesting regularities", there are a few unsupervised learning tasks, including

- *clustering*, which is to find a model $f \in \mathcal{H}$ that divides the training set into clusters such that the clusters satisfy certain intra-cluster similarity and inter-cluster dissimilarity.
- *anomaly detection*, which is to learn a model $f \in \mathcal{H}$ that represents "normal" instances, so that the model can later be used to determine whether a new data instance $\mathbf{x}$ looks normal or anomalous.
- *dimensionality reduction*, which is to find a model $f \in \mathcal{H}$ that represents each instance $\mathbf{x}$ with a lower-dimension feature vector $\mathbf{x}'$, i.e., $|\mathbf{x}'| < |\mathbf{x}|$ while still preserving key properties of $\mathbf{x}$. Key properties can be e.g., intra-cluster similarity and inter-cluster dissimilarity.

1.5.3 Semi-supervised Learning

In addition to the supervised and unsupervised learning, there are other learning tasks such as *semi-supervised learning*, which enables the learning to proceed with a smaller labelled dataset D_1 and a larger unlabelled dataset D_2. This becomes even more important because we have more and more data but it is known that the labelling is usually done by human operators and is very costly.

1.6 Learning Schemes

Learning schemes determine how the learning is conducted. We have mainly two categories:

- *batch learning*, with which the learner is given the training dataset as a batch (i.e. all at once), and
- *online learning*, with which the learner receives the data instances sequentially, and updates the model after processing each new batch of data.

Moreover, we note the existence of a distinction between active and passive learning. For the former, it is generally believed that the learner has some role in determining what data it will be trained. However, for the latter, the learner is simply presented with a training dataset.

1.7 Density Estimation

Density estimation is to construct an estimation of the underlying distribution that generates the dataset D. However, for a high-dimensional problem, an accurate estimation of the full distribution is computationally hard, and it might be sufficient to know, among a (possibly infinite) set of models, which model can lead to the best possibility of generating the dataset D. This section considers a few different ways of obtaining such a best model, according to different requirements.

Without loss of generality, we assume that the set of models is parameterised over a set θ of random variables. For example, a Gaussian distribution is parameterised over its means and variance.

1.7.1 Maximum Likelihood Estimation (MLE)

MLE is a frequentist method. It is to estimate the best model parameters θ that maximise the probability of observing the data from the joint probability distribution. Formally,

$$\theta_{MLE} = \arg\max_{\theta} P(D|\theta) \tag{1.6}$$

The resulting conditional probability $P(D|\theta_{MLE})$ is referred to as the likelihood of observing the data given the model parameters. Note that,

$$\arg\max_{\theta} P(D|\theta) = \arg\max_{\theta} \prod_{(\mathbf{x},y)\in D} P(\mathbf{x}|\theta) \tag{1.7}$$

Considering that the product of probabilities (between 0 to 1) is not numerically stable, we add the log term, i.e.,

$$\begin{aligned}
\theta_{MLE} &= \arg\max_{\theta} \log P(D|\theta) \\
&= \arg\max_{\theta} \log \prod_{(\mathbf{x},y)\in D} P(\mathbf{x}|\theta) \\
&= \arg\max_{\theta} \sum_{(\mathbf{x},y)\in D} \log P(\mathbf{x}|\theta)
\end{aligned} \tag{1.8}$$

1.7.2   Maximum A Posteriori (MAP) Queries

MAP is a Bayesian method. It to estimate the best model parameters θ that explain an observed dataset. MAP query is also called MPE (Most Probable Explanation), because each setting of θ can be seen as an explanation and MAP is to compute the most likely explanation. Formally,

$$\theta_{MAP} = \arg\max_{\theta} P(\theta|D) = \arg\max_{\theta} P(D|\theta)P(\theta) \tag{1.9}$$

As we can see that, the MAP computation involves the calculation of a conditional probability of observing the data given a model, weighted by a prior probability or belief about the model. The only difference between MLE and MAP is on the prior distribution $P(\theta)$, and if $P(\theta)$ is uniform, they are exactly the same.

Similar as MLE, we may add the log term and make Eq. (1.9) into:

$$\begin{aligned} \theta_{MAP} &= \arg\max_{\theta} \log P(D|\theta) + \log P(\theta) \\ &= \arg\max_{\theta} \sum_{(\mathbf{x},y)\in D} \log P(\mathbf{x}|\theta) + \log P(\theta) \end{aligned} \tag{1.10}$$

This expression suggests that the MAP can be seen as the MLE with a regularisation term $\log P(\theta)$.

1.7.3   Expected A Posteriori (EAP)

By Eq. (1.9), we know that θ_{MLE} is to compute the mode of the posterior distribution. This might not reflect the distributional information we need, and we may be interested in e.g., the expected value of θ under D.

$$\mathbb{E}(\theta|D) = \int_{\theta} \theta P(\theta|D)d\theta \tag{1.11}$$

1.8   Ground Truth and Underlying Data Distribution

Once a machine learning model is trained, it is desirable to understand if it performs well enough in terms of some pre-specified properties. These properties are defined according to the requirements specified by the designer or the user of the machine learning model. Therefore, they might be problem specific. Nevertheless, there are typical evaluation methods that are frequently adopted by machine learning practitioners and researchers. In the next two chapters (Chaps. 2 and 3), we review a set of evaluation methods. Before introducing concrete evaluation methods, we explain the ground truth of a learning problem and the underlying data distribution. Without loss of generality, in this chapter, we consider classification task.

In principle, when dealing with a classification task where the data instances have m real-valued features, a machine learning model f is to approximate a target, ground truth function h. That is, the ultimate objective of an evaluation is to understand the gap between $f(\mathbf{x})$ and $h(\mathbf{x})$ for all legitimate instances $\mathbf{x} \in \mathbb{R}^m$. Unfortunately, the input domain $\mathbb{R}^m$ is continuous and may contain an uncountable number of instances. It is unlikely that an evaluation method can exhaustively enumerate all possible instances. Therefore, an evaluation method may need to strike a balance between the exhaustiveness and the cost.

For a problem with m features, we let $X_1, \ldots, X_m$ be the m random variables, each of which corresponds to one of the features. Let

$$P_h(X_1, \ldots, X_m) \tag{1.12}$$

be the underlying data distribution of h, such that each instance $\mathbf{x} = (x_1, \ldots, x_m)$ has a probability density $P_h(\mathbf{x})$. We have that

$$\int_{\mathbf{x} \in \mathbb{R}^m} P_h(\mathbf{x}) = 1 \tag{1.13}$$

For a real world problem, $P_h(\cdot)$ may follow a highly non-linear distribution and it is possible that there are many $\mathbf{x} \in \mathbb{R}^m$ that $P_h(\mathbf{x}) = 0$.

The datasets we are dealing with—such as training, test, and validation datasets—are all assumed to be obtained by sampling from this distribution.

Chapter 2
Model Evaluation Methods

This chapter introduces several typical model evaluation methods that have been widely applied to various practical applications. Model evaluation is traditionally an integral part of the machine learning model development process. It uses statistical methods to help determine the best machine learning model for a given dataset, and help understand how well the machine learning model will perform in the future. All the evaluations are dependent on the training and test datasets that are collected prior to the model development process.

2.1 Test Accuracy and Error

Given the input of a machine learning model follows an unknown distribution P_h, it is straightforward that we may estimate how good the model is by using a set of data instances sampled from the distribution. Test accuracy uses a set of test instances D_{test}, or test dataset, to evaluate the model f by letting

$$Acc(f, D_{test}) = \frac{1}{|D_{test}|} \sum_{(\mathbf{x}, y) \in D_{test}} (1 - \mathbf{I}(f(x), y)) \tag{2.1}$$

where $\mathbf{I}(\cdot, \cdot)$ denotes the 0-1 loss, i.e., $\mathbf{I}(y_1, y_2) = 0$ when $y_1 = y_2$, and $\mathbf{I}(y_1, y_2) = 1$ otherwise. Moreover, the test error is

$$Err(f, D_{test}) = 1 - Acc(f, D_{test}) \tag{2.2}$$

Similar as the above, we can define training accuracy and training error by replacing D_{test} with D_{train}.

© The Author(s), under exclusive license to Springer Nature Singapore Pte Ltd. 2023
X. Huang et al., *Machine Learning Safety*, Artificial Intelligence: Foundations,
Theory, and Algorithms, https://doi.org/10.1007/978-981-19-6814-3_2

2.2   Accuracy w.r.t. Training Set Size

It is interesting to understand, for different machine learning models and different datasets, how the test accuracy changes with respect to the size of the training dataset. Figure 2.1 presents a comparison between decision tree and logistic regression over the California housing dataset. We can see that, in this example, the accuracy of the decision tree increases almost linearly with respect to the size of the training dataset, while the logistic regression increases quickly when the size of the training dataset is small but converges (without a significant increase) afterwards.

2.2.1   How to Plot the Curve?

The algorithm proceeds by following Algorithm 1. For each sample size s, it collects a set of k data into the list acc, such that each datum represents a test accuracy of a machine learning model trained over s instances that are randomly sampled from the training dataset. Then, for each number s on the x-axis, we plot a bar of (mean, variance) over the k data.

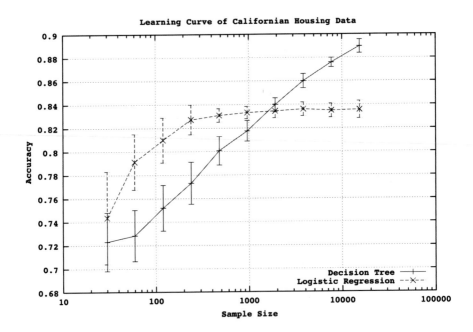

Fig. 2.1 Accuracy w.r.t. training set size [134]

Algorithm 1: $ConstructLearaningCurve(D_{train}, D_{test})$, where D_{train} is a set of training instances and D_{test} is a set of test instances

---

1 **for** *each sample size s on learning curve* **do**
2 | *counter* $= 0$
3 | *acc* $= []$
4 | **while** *counter* $< k$ **do**
5 | | randomly select *s* instances from D_{train}
6 | | learn a model *f*
7 | | evaluate the model *f* on test set D_{test} to determine accuracy *a*
8 | | *acc* $= acc.append(a)$
9 | | *counter* $= counter + 1$
10 | **end**
11 | plot (*s*, average accuracy and error bar) over *acc*
12 **end**

2.3 Multiple Training/Test Partitions

For a real-world application, we may not have enough data to make sufficiently large training and test datasets. In this case, the resulting model may be sensitive to the sizes of the datasets. Specifically, a larger test dataset gives us a more reliable estimate of accuracy (i.e., a lower variance estimate), but a larger training dataset will be more representative of how much data we actually have for the learning process.

In such cases, a single training dataset does not tell us how sensitive accuracy is to a particular training and test split, and we may consider using multiple training/test partitions to evaluate. In the following, we consider several approaches that utilise multiple training/test partitions.

2.3.1 Random Resampling

We can address the issue by repeatedly randomly partitioning the available data D into training dataset D_{train} and test dataset D_{test}. When randomly selecting training datasets, we may want to ensure that class proportions are maintained in each selected dataset.

2.3.2 Cross Validation

The idea is to partition the data into k subsets, and iteratively leaves one out for test and uses the data in the remaining $k - 1$ subsets for training. The resulting accuracy is the average over all the iterations. In general, cross validation makes efficient use of the available data for testing. In practice, 10-fold cross validation is common, but smaller values of k are often used when learning takes a lot of time.

2.4 Confusion Matrix

Up to now, we have some statistical measurements on roughly how good a model is. However, we have not been able to take a look at what types of mistakes the model makes. To this end, a confusion matrix is often used. The confusion matrix is a matrix where each row represents the number of instances in a *predicted* class, while each column represents the number of instances in an *actual* class (or vice versa). Therefore, those numbers on the main diagonal represent the number of instances whose predictive and true labels are the same, and those numbers not on the main diagonal represent mis-classifications.

Example 2.1 Equation (2.4) presents a confusion matrix for the **digits** dataset over 360 test instances, for a Naive Bayes classifier we trained. We can see that, only four instances are mis-classified, two of them are supposed to be labelled as 1 but classified as 2, one of them is supposed to be 6 but classified as 8, and one of them is supposed to be 9 but classified as 5.

$$
\begin{array}{c}
\begin{array}{cccccccccc}
y{=}0 \\ y{=}1 \\ y{=}2 \\ y{=}3 \\ y{=}4 \\ y{=}5 \\ y{=}6 \\ y{=}7 \\ y{=}8 \\ y{=}9
\end{array}
\left(
\begin{array}{cccccccccc}
28 & 0 & 0 & 0 & 0 & 0 & 0 & 0 & 0 & 0 \\
0 & 41 & 0 & 0 & 0 & 0 & 0 & 0 & 0 & 0 \\
0 & 0 & 34 & 0 & 0 & 0 & 0 & 0 & 0 & 0 \\
0 & 0 & 0 & 42 & 0 & 0 & 0 & 0 & 0 & 0 \\
0 & 0 & 0 & 0 & 46 & 0 & 0 & 0 & 0 & 0 \\
0 & 0 & 0 & 0 & 0 & 24 & 0 & 0 & 0 & 1 \\
0 & 0 & 0 & 0 & 0 & 0 & 39 & 0 & 0 & 0 \\
0 & 0 & 0 & 0 & 0 & 0 & 0 & 32 & 0 & 0 \\
0 & 2 & 0 & 0 & 0 & 0 & 1 & 0 & 36 & 0 \\
0 & 0 & 0 & 0 & 0 & 0 & 0 & 0 & 0 & 34
\end{array}
\right) \\
\ \ \ y{=}0\ y{=}1\ y{=}2\ y{=}3\ y{=}4\ y{=}5\ y{=}6\ y{=}7\ y{=}8\ y{=}9
\end{array}
\tag{2.3}
$$

2.4.1 Binary Classification Problem

Consider a binary classification problem. Without loss of generality, we assume that the two classes are *positive* and *negative*, respectively. Then, we have the following confusion matrix

$$
\begin{array}{c}
\begin{array}{c}
y = \textbf{positive} \\ y = \textbf{negative}
\end{array}
\left(
\begin{array}{cc}
\text{true positives (TP)} & \text{false positives (FP)} \\
\text{false negatives (FN)} & \text{true negatives (TN)}
\end{array}
\right) \\
\quad\ \ \ y = \textbf{positive} \qquad\qquad y = \textbf{negative}
\end{array}
\tag{2.4}
$$

where TP denotes the number of true positives, FP the number of false positives, TN the number of true negatives, and FN the number of false negatives. Note that, $|D_{test}| = TP + FP + TN + FP$

We remark that, in this case,

$$Acc(f, D_{test}) = \frac{TP + TN}{|D_{test}|} \text{ and } Err(f, D_{test}) = \frac{FP + FN}{|D_{test}|} \tag{2.5}$$

2.4.2 Other Accuracy Metrics

Is accuracy an adequate measure of predictive performance? Probably not. For example, when there is a large class negative skew, a high accuracy may be misleading.

Example 2.2 For a dataset of 1000 instances, 97% of them are supposed to be negative. In this case, a 98% accuracy may simply be the case that the classifier classifies all negative instances correctly but 2/3 of the positive instances wrongly. This is undesirable for e.g., medical diagnosis, where most of the cases are negative but a true negative may lead to a serious consequence.

To deal with the problem given in Example 2.2, we may consider

$$\text{true positive rate (recall)} = \frac{TP}{TP + FN} \tag{2.6}$$

which focuses on instances whose ground truth are positive. The greater the true positive is, the better the classifier.

Example 2.3 The recall of the case in Example 2.2 is 1/98, which means that the classifier does not perform well with respect to recall.

Similarly, we may consider

$$\text{false positive rate} = \frac{FP}{TN + FP} \tag{2.7}$$

which focuses on instances whose ground truth are negative. However, opposite to the case of true positive rate, the smaller the false positive is, the better the classifier. Moreover, we may be interested in

$$\text{positive predictive value (precision)} = \frac{TP}{TP + FP} \tag{2.8}$$

which focuses on those instances whose predictive values are positive.

Example 2.4 The false positive rate of the case in Example 2.2 is 0, and the precision is 1. The classifier performs well in these two metrics.

2.5 Receiver Operating Characteristic (ROC) Curve

A Receiver Operating Characteristic (ROC) curve plots the true positive rate (Eq. (2.6)) vs. the false positive rate (Eq. (2.7)) when a threshold on the confidence of an instance being positive is varied. Figure 2.2 presents an ROC curve on the **iris** dataset and a logistic classifier.

2.5.1 How to Plot ROC Curve?

First of all, we can get a function **calculate_TP_FP_rate(y_test, y_test_preds)** to compute the TP rate and FP rate given the ground truth labels **y_test** and the predicted labels **y_test_preds**. Then, the Algorithm 2 enables the computation of a set of (TP rate, FP rate) by working with a set of probability thresholds. The key is to compute predicted labels **y_test_preds** based on the predictions **y_predict** and a probability threshold p.

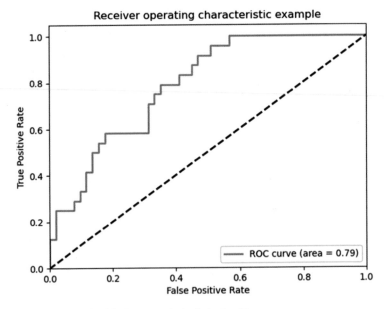

Fig. 2.2 An ROC curve for **iris** dataset and a logistic classifier

---

Algorithm 2: *ROC-curve*(y_test,y_predict), where y_test is the vector of ground truth label, and y_predict is the vector of predictions

---

1 probability_thresholds $= np.linspace(0, 1, num = 100)$
2 **for** *p in probability_thresholds* **do**
3   y_test_preds = []
4   **for** *prob in y_predict* **do**
5   **if** *prob > p* **then**
6   y_test_preds.append(1)
7   **else**
8   y_test_preds.append(0)
9   **end**
10   **end**
11   tp_rate, fp_rate = calculate_TP_FP_rate(y_test, y_test_preds)
12   tp_rates.append(tp_rate)
13   fp_rates.append(fp_rate)
14 **end**
15 **return** tp_rates, fp_rates

2.5.2 How to Read the ROC Curves?

A skilful model will assign a higher probability to a randomly chosen real positive occurrence than a negative occurrence on average. This is what we mean when we say that the model has skill. Generally, skilful models are represented by curves that bow up to the top left of the plot. A model with no skill is represented at the point $(0.5, 0.5)$. A model with no skill at each threshold is represented by a diagonal line from the bottom left of the plot to the top right and has an area under curve (AUC) of 0.5. A model with perfect skill is represented at a point $(0,1)$. A model with perfect skill is represented by a line that travels from the bottom left of the plot to the top left and then across the top to the top right. In general, a greater AUC suggests a better model.

2.6 Precision/Recall (PR) Curves

A precision/recall curve plots the precision (Eq. (2.8)) vs. recall (Eq. (2.6)) (TP-rate) when a threshold on the confidence of an instance being positive is varied. An algorithm can be obtained by adapting Algorithm 2 and we ignore it here. Similar to the ROC curve, a greater area under curve (AUC) suggests a better model.

Chapter 3
Safety and Security Properties

While the model evaluation methods in Chap. 2 can provide certain insights into the quality of machine learning models, their evaluations completely rely on the training and test datasets, without considering other factors that might potentially compromise the performance of a machine learning model. Actually, safety risks may appear at any stage of the lifecycle of a machine learning model. In recent years, the discussion on the potential risks of machine learning models has been very intense, and many risks have been discovered, even though the models have achieved excellent performance with respect to the traditional model evaluation methods. This chapter will discuss a set of safety and security properties.

Simply speaking, a learned model f is to approximate a target function h. Therefore, the erroneous behaviour of f exists when it is inconsistent with h. We use $f_j(\mathbf{x})$ to denote its j-th element of $f(\mathbf{x})$. Then, we have the following definition for the classification task.

Definition 3.1 (Erroneous Behavior of a Classifier) Given a (trained) classifier $f : \mathbb{R}^n \rightarrow \mathbb{R}^k$, a target function $h : \mathbb{R}^n \rightarrow \mathbb{R}^k$, an erroneous behavior of the classifier f is exhibited by a legitimate input $\mathbf{x} \in \mathbb{R}^n$ such that

$$\arg\max_j f_j(\mathbf{x}) \neq \arg\max_j h_j(\mathbf{x}) \tag{3.1}$$

Intuitively, an erroneous behaviour is witnessed by the existence of an input $\mathbf{x}$ on which the classifier and the target function return a different label. Note that, the legitimate input $\mathbf{x}$ can be any input in the input domain $\mathbb{R}^n$ and does not have to be a training instance.

In the following, we discuss several classes of properties that might affect the safe use of machine learning models in real-world safety-critical applications.

X. Huang et al., *Machine Learning Safety*, Artificial Intelligence: Foundations, Theory, and Algorithms, https://doi.org/10.1007/978-981-19-6814-3_3

3.1 Generalisation Error

One of the key successes of machine learning is that it is able to work with unseen data, i.e., data instances that are not within the training dataset. It is meaningful to understand how good a model is on unseen data. Given an instance $\mathbf{x}$, we use a loss function to measure the discrepancy between the true label y and the model's predicted label $f(\mathbf{x})$, written as $\mathcal{L}(y, f(\mathbf{x}))$.

Given a training dataset D_{train} sampled i.i.d. from the underlying distribution $\mathcal{D}$, the model f's empirical loss on D_{train}, or train loss, is

$$\mathcal{L}_{emp}(f, D_{train}) \stackrel{\text{def}}{=} \frac{1}{|D_{train}|} \sum_{(\mathbf{x},y) \in D_{train}} \mathcal{L}(y, f(\mathbf{x})) \tag{3.2}$$

while the expected loss is

$$\mathcal{L}_{exp}(f, \mathcal{D}) \stackrel{\text{def}}{=} \mathbb{E}_{(\mathbf{x},y) \in \mathcal{D}} \mathcal{L}(y, f(\mathbf{x})) \tag{3.3}$$

Their gap

$$GE(f, D_{train}, \mathcal{D}) = |\mathcal{L}_{emp}(f, D_{train}) - \mathcal{L}_{exp}(f, \mathcal{D})| \tag{3.4}$$

is called generalisation error. The empirical test loss

$$\mathcal{L}_{emp}(f, D_{test}) \stackrel{\text{def}}{=} \frac{1}{|D_{test}|} \sum_{(\mathbf{x},y) \in D_{test}} \mathcal{L}(y, f(\mathbf{x})) \tag{3.5}$$

is often used to approximate the expected loss $\mathcal{L}_{exp}(f, \mathcal{D})$, since the underlying distribution $\mathcal{D}$ is unknown to the learning algorithm. Therefore, $GE(f, D_{train}, \mathcal{D})$ can be approximated with

$$GE(f, D_{train}, D_{test}) = |\mathcal{L}_{emp}(f, D_{train}) - \mathcal{L}_{emp}(f, D_{test})| \tag{3.6}$$

Recall that, the test error $Err(f, D_{test})$ (Eq. (2.2)) is an empirical test loss when the loss function $\mathcal{L}$ is the 0-1 loss.

Generalisation error is related to the well-known overfitting problem of machine learning algorithms. A machine learning model is overfitted if it performs well on training data instances but badly on test data instances. Usually, a large generalisation error indicates overfitting. Moreover, generalisation error is also related to the representativeness of training and test datasets. For a high-dimensional problem, if the training and test datasets do not contain a sufficiently large amount of data instances, the estimated generalisation error will not be accurate. An inaccurate estimation will lead to a bad judgement on the quality of the machine learning model over future inputs and may lead to safety implications.

As will be discussed in Chap. 12.2, there are other factors that may affect the generalisation ability of machine learning models, including e.g., the hyper-parameters, the model structure, and the learnable parameters.

3.2 Uncertainty

Machine learning can be seen as an approach of implementing inductive reasoning, because it generalises from a set of observations and obtains a model that can serve as a general principle. However, a learned model will not be provably correct—as discussed in Sect. 3.1 and Chap. 2, a trained model f may output incorrect decisions. Uncertainty quantification in machine learning is to quantitatively measure the uncertainty in the decisions made by a machine learning model.

Other than the inherent uncertainty from the above-mentioned inductive infer-ence, considering the development process of machine learning, there are other uncertainties such as the incorrect model assumption, nonoptimal hyper-parameters, and noisy and imprecise data. A solid representation of uncertainty may contribute well to the trustworthiness of AI, as discussed in Sect. 16.5.

There are two inherently different sources of uncertainty: aleatoric and epistemic uncertainty. *Aleatoric uncertainty* refers to the variability in the outcome of an experiment due to an inherently random effect. For example, in a coin-flipping experiment, the outcome of the experiment is inherently random. The inherent stochastic nature suggests that aleatoric uncertainty cannot be reduced even with the optimal model. On the other hand, *epistemic uncertainty* refers to the lack of knowledge. The decision-maker may not have the full information about the problem due to the ignorance. For example, on seeing the observations 0,0,1 of a series of coin-flipping experiments, it is unclear whether we should model the coin as a fair one or not. As opposed to the aleatoric uncertainty, epistemic uncertainty can be reduced or eliminated if provided with sufficient information, such as infinite observations in the coin-flipping experiments. In other words, we may roughly refer aleatoric uncertainty as the irreducible part of the total uncertainty and epistemic uncertainty as the reducible part of the total uncertainty.

Sources of Uncertainty for Classification

First of all, the classification problem $f(\mathbf{x})$ is actually to find a conditional probability $p(y|\mathbf{x})$, which returns a probability value for every class $y \in C$ when observing an instance $\mathbf{x}$. However, due to the above-mentioned uncertainties, such a conditional probability is not fixed, i.e., there may be multiple, or even infinite, probabilities $p(y|\mathbf{x})$ when given $\mathbf{x}$. Based on this, we can have a Bayes predictor h^*

that computes

$$h^*(\mathbf{x}) \triangleq \arg\min_{\hat{y}} \int_y \mathcal{L}(y, \hat{y}) dp(y|\mathbf{x}) \tag{3.7}$$

for every input instance $\mathbf{x}$. We can see that, the Bayes predictor h^* considers all possibilities of $p(y|\mathbf{x})$ and intends to get the best prediction on average. We remark that, the uncertainty in $p(y|\mathbf{x})$ is aleatoric uncertainty, as it is inherently random.

Moreover, given a hypothesis class $\mathcal{H}$ and Eq. (3.3), we may find the best classifier among $\mathcal{H}$, i.e.,

$$f^* \triangleq \arg\min_{f \in \mathcal{H}} \mathcal{L}_{exp}(f, \mathcal{D}) \tag{3.8}$$

The gap between f^* and h^* is from the uncertainty about the right type of model to construct and fit. We call such uncertainty as *model uncertainty*.

Finally, training algorithms are locally optimal, instead of globally optimal, which means that we may not be able to achieve f^* in the end, but rather a model $\hat{f}$. The gap between $\hat{f}$ and f^* is from an uncertainty we call *approximation uncertainty*.

Both model uncertainty and approximation uncertainty are epistemic uncertainty, because they are due to the lack of knowledge and can—in theory—be reduced. Also, the above three uncertainties (uncertainty in $p(y|\mathbf{x})$, model uncertainty, approximation uncertainty) correspond to the three errors (Bayes error, model error, approximation error) to be discussed in Sect. 16.3.

3.3 Robustness and Adversarial Examples

Adversarial examples [165] represent another class of erroneous behaviours that also introduce safety implications. Here, we take the name "adversarial example" due to historical reasons. Actually, as suggested in the below definition, it represents a mis-match of the decisions made by a human and by a machine learning model, and does not necessarily involve an adversary.

Intuitively, as shown in Fig. 3.1, it is possible that, given an instance $\mathbf{x}$, it is classified correctly, i.e., $y = \hat{f}(\mathbf{x})$, but a small perturbation on $\mathbf{x}$ (such as the one-pixel change as in Fig. 3.1) may lead to a change of classification, i.e., $\hat{f}(\mathbf{x} + \epsilon) \neq \hat{f}(\mathbf{x})$. Such misclassification may have serious security implications, as the traffic light example in Fig. 3.1.

For the **iris** dataset, if we create a new data instance, indexed as 151 in the below Fig. 3.2a and b, such that the only difference with the one indexed as 150 is the petal width, from 1.8 to 1.6. A well trained decision tree classifier may classify the new instance as a different class.

DL Classification: Green Light   **DL Classification: Red Light**

Fig. 3.1 By changing one pixel in a "Green-Light" image, a state-of-the-art deep learning (DL) image classifier misclassifies it as "Red-Light" [181]

a   **b**

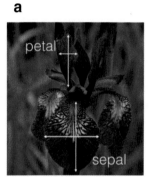

index	Sepal Length	Sepal Width	Petal Length	Petal Width	Class Label
1	5.1	3.5	1.4	0.2	iris setosa
2	4.9	3.0	1.4	0.2	iris setosa
...					
50	6.4	3.5	4.5	1.2	iris versicolor
...					
150	5.9	3.0	5.1	1.8	iris virginica
151	5.9	3.0	5.1	1.6	iris versicolor

Fig. 3.2 (a) An iris flower. (b) Iris dataset

Definition 3.2 (Adversarial Example) Let s_1 be the number of input features and $s_K = |C|$ be the number of classes. Given a (trained) machine learning classifier $f : \mathbb{R}^{s_1} \to \mathbb{R}^{s_K}$, a human decision oracle $h : \mathbb{R}^{s_1} \to \mathbb{R}^{s_K}$, and a legitimate input $\mathbf{x}$, we write $\hat{h}(\mathbf{x})$ for the ground truth label assigned by the human decision oracle h, i.e., $\hat{h}(\mathbf{x}) = \max_j h_j(\mathbf{x})$. Assume that $\hat{f}(\mathbf{x}) = \hat{h}(\mathbf{x})$, i.e., $\mathbf{x}$ is a correctly labelled data instance. Then, the existence of an adversarial example to f and $\mathbf{x}$ is defined as:

$$\exists \mathbf{x}' : \hat{h}(\mathbf{x}') = \hat{h}(\mathbf{x})$$
$$\land \hat{f}(\mathbf{x}') \neq \hat{f}(\mathbf{x}) \tag{3.9}$$

We recall that, $\hat{f}(\mathbf{x}) = \arg\max_j f_j(\hat{\mathbf{x}})$ and $\hat{f}(\mathbf{x}') = \arg\max_j f_j(\mathbf{x})$ are predictive labels of $\mathbf{x}$ and $\mathbf{x}'$, respectively.

Intuitively, $\mathbf{x}$ is an input on which the classifier and a human user have the same classification label and, based on this, an adversarial example is another input $\mathbf{x}'$ that is classified differently than $\mathbf{x}$ by classifier f (i.e., $\hat{f}(\mathbf{x}') \neq \hat{f}(\mathbf{x})$), even the human user believes that they should be the same (i.e., $\hat{h}(\mathbf{x}') = \hat{h}(\mathbf{x})$).

However, in practice human decision oracle (i.e., $\hat{h}(\mathbf{x})$) is hard to obtain, so usually we adopt a certain distance metric (e.g., L_p-norm metric or other types of metrics) to approximate the human decision in practise, for example, Eq. (3.9) can be specifically defined as:

$$\exists \mathbf{x}' : \ ||\mathbf{x} - \mathbf{x}'||_p \leq d \\ \wedge \hat{f}(\mathbf{x}') \neq \hat{f}(\mathbf{x}) \tag{3.10}$$

where $p \in \mathbb{N}$, $p \geq 1$, $d \in \mathbb{R}$, and $d > 0$ is a small positive number, $|| \cdot ||_p$ is p-norm as defined in Sect. B.3.

3.3.1   Measurement of Adversarial Examples

Definition 3.2 explains the adversarial example, but given that the human decision oracle is hard to define and there may be multiple adversarial examples satisfying the conditions, there needs to be some quantifiable measurement by which we may decide that some adversarial examples are more interesting than others.

Definition 3.3 (Quality of Adversarial Examples) An adversarial example is usually measured from the following two aspects:

- magnitude of perturbation, i.e., $||\mathbf{x} - \mathbf{x}'||$, where $|| \cdot ||$ is a norm distance such as those introduced in Sect. B.3,
- difference of prediction confidence before and after the perturbation, i.e., $|f_y(\mathbf{x}) - f_y(\mathbf{x}')|$.

Therefore, instead of concerning any adversarial example satisfying Definition 3.2, we are interested in the following optimisation problem, for some instance $(\mathbf{x}, y) \in \mathcal{D}$,

$$\begin{aligned} \text{minimise } & ||\mathbf{x} - \mathbf{x}'|| - \lambda |f_y(\mathbf{x}) - f_y(\mathbf{x}')| \\ \text{subject to } & \hat{f}(\mathbf{x}) \neq \hat{f}(\mathbf{x}') \\ & ||\mathbf{x} - \mathbf{x}'|| \leq d \end{aligned} \tag{3.11}$$

where λ is a hyper-parameter balancing two objectives, and d indicates the maximum perturbation that may be considered.

3.3.2 Sources of Adversarial Examples

The adversarial examples are legitimate data instances, except that they are forged by adding perturbations to the correctly labelled data instances. The perturbation may come from different sources. It is possible that there is an *adversarial agent* who deliberately adds carefully crafted perturbation to make the machine learning classifier misclassify. We call it malicious perturbation. On the other hand, the perturbation does not have to be from an adversarial agent. Instead, it may be noise from the environment, such as the white noise of the sensor and the camera. We call it benign perturbation. We may also regard the benign perturbation to be from a *benign agent*.

3.3.3 Robustness

Robustness is a dual concept of adversarial examples. It requires that the decision of a machine learning model is invariant against small perturbations, i.e., the adversarial example does not exist. The following definition is adapted from that of [70].

Definition 3.4 (Local Robustness) Given a machine learning model f with s_1 input features and s_K classes, and an input region $\eta \subseteq [0, 1]^{s_1}$, the (un-targeted) local robustness of f on η is defined as

$$\forall \mathbf{x} \in \eta, \exists l \in [1..s_K], \forall j \in [1..s_K] : f_l(\mathbf{x}) \geq f_j(\mathbf{x}) \tag{3.12}$$

For targeted local robustness of a label $j \in C$, it is defined as

$$\forall \mathbf{x} \in \eta, \exists l \in [1..s_K] : f_l(\mathbf{x}) > f_j(\mathbf{x}) \tag{3.13}$$

Intuitively, local robustness states that all inputs in the region η have the same class label. More specifically, there exists a label l such that, for all inputs $\mathbf{x}$ in the region η, and other labels j, the machine learning model believes that $\mathbf{x}$ is more possible to be in class l than in any class j. Moreover, targeted local robustness means that a specific label j cannot be perturbed for all inputs in η; specifically, all inputs $\mathbf{x}$ in η have a class $l \neq j$, which the machine learning model believes is more possible than the class j. Usually, the region η is defined with respect to an input $\mathbf{x}$ and a norm L_p, as in Definition B.1. If so, it means that all inputs in η have the same class as input $\mathbf{x}$. For targeted local robustness, it is required that none of the inputs in the region η is classified as a given label j.

3.4 Poisoning and Backdoor Attacks

Adversarial examples make the machine learning classifier misclassify. They do not impose any change to the classifier, and only fool a trained classifier by perturbing the input. In this section, we introduce two other safety errors that may require a change to either the training dataset or the training process to make the model misclassify. These two errors are due to the adversary injecting malicious information into the machine learning lifecycle and hence getting a machine learning algorithm to learn something it should not.

3.4.1 Data Poisoning Attack

Data poisoning attack adds malicious data instances into the training dataset, to deliberately control the classification of certain data instances. It can be formalised as a bi-level optimisation as follows. Assume that the attacker intends to force an input $\mathbf{x}_{adv}$ to be predicted as a label y_{adv}. To implement so, the attacker adds a set of poisoned inputs $\mathbf{X}_p$ to the original dataset D_{train}. The question is then to find the optimal $\mathbf{X}_p$. Formally, the optimal $\mathbf{X}_p$ is

$$\mathbf{X}_p^* = \arg\max_{\mathbf{X}_p} \mathcal{L}_{adv}(\mathbf{x}_{adv}, y_{adv}; \mathbf{W}^*(\mathbf{X}_p)) \tag{3.14}$$

where $\mathcal{L}_{adv}$ is a loss function measuring the loss a classifier with parameters $\mathbf{W}^*(\mathbf{X}_p)$ assigns label y_{adv} to $\mathbf{x}_{adv}$, and $\mathbf{W}^*(\mathbf{X}_p)$ are the parameters of the classifier trained on $\mathbf{X}_p \cup D_{train}$. Note that, Eq. (3.14) is a bi-level optimisation because

$$\mathbf{W}^*(\mathbf{X}_p) = \arg\min_{\mathbf{W}} \mathcal{L}_{train}(\mathbf{X}_p \cup D_{train}, \mathbf{y}; \mathbf{W}) \tag{3.15}$$

where $\mathcal{L}_{train}$ is the standard training loss (such as the cross-entropy loss), and $\mathbf{y}$ contains correct labels for both $\mathbf{X}_p$ and D_{train}.

Figure 3.3 gives an example of a data poisoning attack on an image classifier. By adding a set of poisoning images to the set of clean images, it makes the resulting classifier misclassify the target image, whose true label is *Bird*, as *Dog*.

3.4.2 Backdoor Attacks

Given a *triggered* input $\mathbf{x}^\alpha = \mathbf{x} + \Delta$, where Δ is the trigger stamped on a "clean" input $\mathbf{x}$, the predicted label will always be the label y_α that is set by the attacker, regardless of what the input $\mathbf{x}$ might be. In other words, as long as the triggered input $\mathbf{x}^\alpha$ is present, the backdoor model will always classify the input to the attacker's

Clean Images Clean Images Poisons Poisons

Target:
True Class: Bird
Poisoned : Dog

Fig. 3.3 Example of data poisoning attack [65]

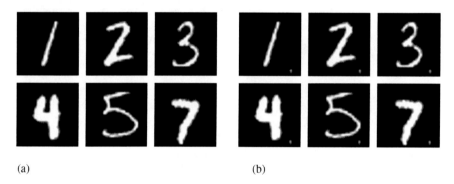

(a) (b)

Fig. 3.4 All MNIST images of handwritten digit with a backdoor trigger (a white patch close to the bottom right of the image) are mis-classified as digit 8 [67]. (**a**) Clean inputs representing different digits. (**b**) Backdoor inputs, all classified as 8

target label (i.e. y_α). However, for "clean" inputs, the backdoor model behaves as the original model without any observable performance reduction. Figure 3.4 presents an example backdoor attack on MNIST dataset. We can see that, with a trigger, any handwritten image will be classified as 8.

To facilitate the backdoor attack, we can consider a data poisoning attack by adding poisoned instances to the training dataset. Alternatively, we can consider a direct modification to the trained model, as will be discussed in Sect. 5.4 for decision tree classifiers.

3.4.3   Success Criteria of Attack

Let f' be the attacked model. To evaluate how successful a poisoning attack is, we suggest the following criteria [67]:

- (Preservation, or P-rule) f' has similar performance as f on a test dataset.

Actually, P-rule suggests that a poisoned model performs similarly on the natural data that are on the same distribution as the training data. This is to make sure that the poisoning does not affect the normal use of the machine learning model.

- (Verifiability, or V-rule) The attacker is able to verify whether the attack has been conducted on the model f'.

V-rule requires that the attacker is able to check whether the poisoning is actually conducted, without e.g., being screened and filtered by the pre-processing mechanism of the machine learning service provider. For backdoor attacks, this is an essential requirement, and easy to verify, as an attacker will know if V-rule holds by querying the model with patched input instances.

- (Stealthiness, or S-rule) It is hard to differentiate f and f'.

S-rule suggests that the poisoning should not be easily detected. Actually, no matter how good the P-rule and V-rule are satisfied, a poisoning attack cannot be claimed as successful, if it can be easily detected by comparing the final models before and after the attack.

3.5   Model Stealing

Machine learning model can be seen as confidential as it might involve commercially sensitive data (such as trained model and training dataset) that might need to be protected. In the next two safety and security issues (Sects. 3.5 and 3.6), we consider the potential of the machine learning model being attacked such that the trained model or the training data is leaked. This may occur when the machine learning model is deployed as e.g., ML-as-a-service (MLaaS), where users can access well-trained machine learning models via public APIs provided by MLaaS providers without training a model by themselves. In practice, such leakages may lead to significant financial loss or privacy loss.

Given a model f, a model stealing agent is to reconstruct another model f'. The reconstruction may have different requirements, for example,

- reconstruct the hyper-parameters,
- reconstruct the model and the trainable parameters, and
- reconstruct another model that is functionally equivalent to f.

Moreover, for different requirements, the attacker may be given different knowledge. We will attacker knowledge in Sect. 3.7.

3.6 Membership Inference and Model Inversion

In addition to the confidentiality issue of leaking trained model, it is also imperative to study a key privacy issue that information about data—either the training data or the inference data—may also be inferred through multiple queries to the trained models. We consider two classes of privacy issues, i.e., membership inference and model inversion.

3.6.1 Membership Inference

Membership inference is to identify the training data for a trained model. Formalised as a decision problem, it is, given a data instance $\mathbf{x}$ and the access to a model f, to determine if the instance $\mathbf{x}$ was in the model's training dataset, i.e., if $\mathbf{x} \in D_{train}$. The access to the model can be either white-box or black-box (will be introduced in Sect. 3.7), depending on the concrete scenarios.

Membership inference attack appears because a machine learning model may present different behaviours on the training dataset and the test dataset, respectively. Machine learning, in particular deep neural networks, is often overparameterised (i.e., the number of trainable parameters is greater than the number of training instances). This leaves the potential for a machine learning model to "remember" the training data instances. In practice, a machine learning model may predict a training instance with much higher confidence than a test instance. Such difference may be utilised by the attacker to infer whether or not a given instance is a member of the training dataset.

3.6.2 Model Inversion

Model inversion is to infer sensitive information (e.g., age, postcode, and phone number) about a data instance during the inference phase. Assume that each data instance includes m features $X_1, \ldots, X_m$. Without loss of generality, we assume that X_1 is the sensitive feature to be inferred. Then, given partial information about a data instance $\mathbf{x}$ (e.g., values of features $X_2, \ldots, X_n$) and its predictive label y by a machine learning model f, it is to infer the value for sensitive feature X_1. Figure 3.5 gives an illustrative example on model inversion on images.

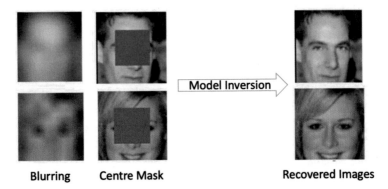

Fig. 3.5 An illustrative example on model inversion attack [195]. The model inversion attack is able to recover the sensitive part of the images

Another problem formulation of model inversion is to reconstruct an instance **x** from available information such as a predictive label $\hat{y}$ or a predictive output probability $f(\hat{\mathbf{x}})$.

3.7 Discussion: Attacker Knowledge and Attack Occasions

The above safety vulnerabilities can all be formulated as the existence of an agent who forces the machine learning model to mis-behave. Figure 3.6 presents an overall picture of the occasions of different attacks in the development cycle of a machine learning model. Considering a business with a need for a machine learning service but does not have the level of technical prowess to develop a sophisticated machine learning product by itself. It might outsource the data collection and preparation, model design and training, and user service to different technology companies. While convenient, this might lead to potential risks of being attacked. As can be seen from Fig. 3.6, the attacks that are introduced in this chapter might work on different phases of the entire process.

The agent can be benign or malicious. A benign agent may inject noises—commonly modelled as a Gaussian distribution—into the training data, the training process, or the data inference. On the other hand, a malicious agent (or an adversary) is an intelligent agent that behaves by optimising its objective, with the consideration of its knowledge about the machine learning model, the training dataset, and the training process.

The following is a list of possible knowledge of the adversary:

K1 training dataset
K2 the distribution of the training data
K3 the type of machine learning model, such as decision tree, neural network etc.
K4 trained parameters of the machine learning model

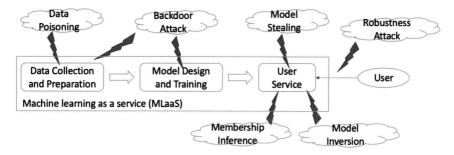

Fig. 3.6 Attack occasion in machine learning as a service (MLaaS)

K5 hyper-parameters of training process
K6 ability of observing the output (predictive label, prediction probability, etc) of an input instance when inference

where **K1–K2** are about the training data, **K3–k4** are about the trained model, **K5** is about the training process, and **K6** is about the inference phase.

According to the knowledge that is available to the adversary, there are roughly two classes of adversaries:

- **black-box adversary**, which is assumed to have the ability of observing the output of an input (i.e., **K6**), but without other knowledge (i.e., **K1–K5**)
- **white-box adversary**, which is assumed to have all the knowledge (i.e., **K1–K6**)

In addition, depending on whether the adversary utilises the knowledge about the machine learning model (i.e., **K3–K4**), we have

- **model specific attack**, where the attack algorithm requires **K3–K4**
- **model agnostic attack**, where the attack algorithm does not require **K3–K4**

It is not hard to see that, a black-box adversary can only use a model agnostic attack, but not vice versa.

Chapter 4
Practice

In this chapter, we will explain how to install an experimental environment based on Python, and present a few examples of basic Python operations and the code for the visualisation of a simple dataset (as in Example 1.4) and the confusion matrix (as introduced in Sect. 2.4).

4.1 Installation of Experimental Environment

First of all, your machine needs to have Python3 or Anaconda installed. Then, to install the python packages that will be used later, you have the following two options.

4.1.1 Use Pip

The first option is to use Pip for installation. First, you need to make sure that python and pip (or pip3) are installed appropriately. You can check this with the 'pip -V' in Windows Commands (cmd) or MacOS Terminal. Then, the installation of software packages is done through the following example:

```
$ pip (or pip3) install matplotlib
```

© The Author(s), under exclusive license to Springer Nature Singapore Pte Ltd. 2023
X. Huang et al., *Machine Learning Safety*, Artificial Intelligence: Foundations,
Theory, and Algorithms, https://doi.org/10.1007/978-981-19-6814-3_4

4.1.2   Create Conda Virtual Environment

You can download the Anaconda through https://www.anaconda.com/products/individual, create an environment and install some necessary software packages:

```
$ conda create --name aisafety
$ conda activate aisafety
$ conda install pandas numpy matplotlib tensorflow scikit-learn \
    pytorch torchvision
```

In addition to the installation of packages, you could install Jupyter notebook through https://jupyter.org/install, for the editing and running of python code via a web browser.

4.1.3   Test Installation

Once installed, you can check whether the installation is successful by running the following commands. Note that, the first command is to activate the Python environment, in which the remaining commands are executed.

```
$ python3
$ import numpy
$ import scipy
$ import sklearn
$ import matplotlib
$ import pandas
$ exit()
```

Once the installation is successful, create a new file **lab1.py** and type in the following lines:

```
import numpy as np

x = np.arange(100)
y = np.array(5)

z=x+y
```

Then, you can use the following command to check the result:

```
$ python (or python3) lab1.py
```

4.2 Basic Python Operations

In the following, we provide a few exercises for basic Python operations. Please complete them sequentially.

1. Using array indexing to give the last ten values of z

```
xLastTen = x[90:] # Or x[-10:]
```

2. Update the code so that x goes from 0 to 1000 in steps of 10

```
xUpdate = np.arange(0, 1000, 10)
```

3. Take dot product between x and x

```
xDotProduct = x.dot(x)
```

4. Take (*) product between x and x

```
xAsteriskProduct = x * x
```

5. Reshape x so that it is no longer a 100 value array but a 10x10 matrix

```
xReshape = xUpdate.reshape((10, 10))
```

6. Multiply the first row by 1, the second by 2, the third by 3 and so on

```
yNew = np.arange(1,11)
zNew = xReshape * yNew[:, np.newaxis]
print(zNew)
```

7. Using the matplotlib library to plot each row of this matrix as a single series on the same graph

```
for i in range(10):
    plt.plot(zNew[i])
plt.show()
```

8. Using the matplotlib library to plot each row of this matrix as a single series on separate sub-plots of the same figure and save this figure as figure1.png

```
for i in range(10):
    ax = plt.subplot(5, 2, i + 1)
    plt.plot(zNew[i])
plt.savefig('figure1.png')
plt.show()
```

4.3 Visualising a Synthetic Dataset

Below is the code for the visualisation of the synthetic dataset as given in Example 1.4.

```
import numpy as np
import matplotlib.pyplot as plt

xdata = 7 * np.random.random(100)
ydata = np.sin(xdata) + 0.25 * np.random.random(100)
zdata = np.exp(xdata) + 0.25 * np.random.random(100)

fig = plt.figure(figsize=(9, 6))
# Create 3D container
ax = plt.axes(projection = '3d')
# Visualize 3D scatter plot
ax.scatter3D(xdata, ydata, zdata)
# Give labels
ax.set_xlabel('x')
ax.set_ylabel('y')
ax.set_zlabel('z')
# Save figure
plt.savefig('3d_scatter.png', dpi = 300);
```

To run this code, you need to activate the environment you have created earlier, with the following command:

```
$ conda activate aisafety
$ python3 4.3.txt
```

This will be needed for all future practicals.

4.4   Confusion Matrix

We train a simple perceptron model and output the confusion matrix.

```
from sklearn import datasets
dataset = datasets.load_digits()
X = dataset.data
y = dataset.target

print("===== Get Basic Information ======")
observations = len(X)
features = len(dataset.feature_names)
classes = len(dataset.target_names)
print("Number of Observations: " + str(observations))
print("Number of Features: " + str(features))
print("Number of Classes: " + str(classes))

print("===== Split Dataset ======")
from sklearn.model_selection import train_test_split
X_train, X_test, y_train, y_test = train_test_split(X, y,
    test_size=0.20)

print("===== Model Training ======")
from sklearn.linear_model import Perceptron
```

```
20 clf = Perceptron(tol=1e-3, random_state=0)
21 clf.fit(X_train, y_train)
22
23 print("===== Model Prediction ======")
24 print("Labels of all instances:\n%s"%y_test)
25 y_pred = clf.predict(X_test)
26 print("Predictive outputs of all instances:\n%s"%y_pred)
27
28 print("===== Confusion Matrix ======")
29 from sklearn.metrics import classification_report,
     confusion_matrix
30 print("Confusion Matrix:\n%s"%confusion_matrix(y_test, y_pred))
31 print("Classification Report:\n%s"%classification_report(y_test,
     y_pred))
```

Exercises

Question 1 Give an example for a supervised learning problem, an unsupervised learning problem, and a semi-supervised learning problem; □

Question 2 Give an example of a supervised learning problem that is a classification task; □

Question 3 Give an example of a supervised learning problem that is a regression task; □

Question 4 Give an example of a clustering problem; □

Question 5 For all the above problems, figure out the features and labels for them; □

Question 6 Write a program to output the following information:

1. How many samples are in the **iris** dataset;
2. How many features are given for each sample in the **iris** dataset?
3. What is the value range for each feature?

□

Question 7 According to Table 4.1 about two random variables $Intelligence$ and $Grade$, please compute the values $P(Grade = B \mid Intelligence = Low)$ and $MAP(Grade)$.

Question 8 Consider a joint distribution table as in Table 4.2, can you compute the following expressions:

Table 4.1 Joint probability for student grade and intelligence

	Intelligence = Low	Intelligence = High
Grade = A	0.07	0.18
Grade = B	0.28	0.09
Grade = C	0.3	0.08

Table 4.2 Joint distribution table

	B=1	B=2	B=3	B=4
A=1	0.12	0.18	0.24	0.06
A=2	0.06	0.09	0.12	0.03
A=3	0.02	0.03	0.04	0.01

Table 4.3 Joint distribution table

	B=1	B=2	B=3	B=4
A=1	0.12	0.18	0.24	0.02
A=2	0.06	0.09	0.12	0.03
A=3	0.06	0.03	0.04	0.01

- $P(A=1)=0.6$
- $P(A=2)=0.3$
- $P(B=3)=0.4$
- $P(B=4)=0.1$
- $P(A=1|B=2)=0.6$
- $P(B=3|A=3)=0.4$
- $MAP(A|B=2)=1$
- $MAP(B|A=2)=3$
- $MAP(A)=1$
- $MAP(B)=3$

□

Question 9 Consider a joint distribution table as in Table 4.3, can you compute the following expressions:

- $P(A=1)=0.56$
- $P(A=2)=0.3$
- $P(B=3)=0.4$
- $P(B=4)=0.06$
- $P(A=1|B=2)=0.6$
- $P(B=3|A=3)=2/7$
- $MAP(A|B=2)=1$
- $MAP(B|A=2)=3$
- $MAP(A)=1$
- $MAP(B)=3$

□

Question 10 Write a program to implement

- ROC curve
- PR curve

□

Question 11 Compare a few training/test splits (0.9/0.1, 0.8/0.2, 0.7/0.3) and check their differences on training and test accuracy. □

Question 12 Compare a few training/test splits (0.9/0.1, 0.8/0.2, 0.7/0.3) and check their differences on confusion matrix.   □

Question 13 Find a data poisoning strategy to make the trained model mis-classify on a given training input.   □

Question 14 Read the literature to understand the state-of-the-art for backdoor attacks.   □

Part II
Safety Threats

This part will exploit the machine learning models to understand the threats to their safety and security. We will focus on not only traditional machine learning models but also deep learning. Traditional machine learning models to be considered include decision trees, k-nearest neighbour, linear regression, and Naive Bayes. For deep learning, we will consider feedforward neural networks, which include convolutional neural networks. Besides, two key concepts will be introduced: gradient descent, and loss functions, which are useful for both traditional machine learning and deep learning on not only their training algorithms but also their related attack, defence, and verification algorithms.

For each machine learning model, instead of moving directly into the discussion on its safety threats, we will first explain its basic knowledge, including the structure of the model and its training algorithm. This is followed by presenting safety threats through various algorithms to exploit different vulnerabilities. The safety threats to be discussed are with respect to the properties we discussed in Chap. 3. These algorithms will help the readers understand the safety and security issues and hopefully inspire new, better algorithms.

For every algorithm on some machine learning model, we will often discuss it by considering a few aspects of the algorithms as follows. First of all, we need to know if the algorithm is *sound*, i.e., whether the safety vulnerability discovered by the algorithm is actually an issue of the model. This is trying to understand if the algorithm may generate false alarms. The second is to discuss if the algorithm is *complete*, i.e., whether a report of failure in finding safety vulnerability by the algorithm actually suggests the missing of safety vulnerability. The third aspect to be considered is the information required by the algorithm, i.e., if it is black-box or white-box. The fact that an algorithm is white-box usually suggests that it is dedicated to a certain machine learning model and may not be transferable to other machine learning models. On the other hand, a black-box algorithm usually suggests that the algorithm is applicable to different machine learning models. Moreover, for algorithms, we may want to know their computational complexity, as an indicator of how hard a vulnerability can be discovered.

Chapter 5
Decision Tree

Decision tree is one of the simplest, yet popular, machine learning algorithms. It has a very long history of research and application, and has many variants. This chapter will present a training algorithm for decision trees and discuss several algorithms to identify the safety and security risks of a trained decision tree or its training algorithm.

Figure 5.1 shows a decision tree for the **iris** dataset. We can see that every internal node, including the root node, is attached with a condition, such that the satisfiability of the condition leads to

As an example of decision tree, we can convert any Boolean formula into a decision tree. For example, Fig. 5.2 presents decision trees for the formulas $X_2 \wedge X_5$ and $X_2 \vee X_5$, respectively. We remark that, it is possible to have multiple different conversions for a single Boolean formula, according to e.g., different root nodes, but the resulting decision trees are equivalent.

5.1 Learning Algorithm

Algorithm 3 provides a program sketch of a function $Construct SubTree(D)$ for D a set of training instances. Intuitively, given D, this function will construct and return a tree T_D to classify the training instances in D. The tree construction is a recursive process, i.e., the construction of T_D is completed by the construction of its children $T_{D_1}, \ldots, T_{D_k}$, which are implemented by calling $Construct SubTree(D_i)$ for $i \in \{1, \ldots, k\}$ such that $D = \bigcup_{i=1}^{k} D_i$. Once D satisfies the stopping criteria, a leaf node is constructed and the recursive process terminates by making T_D be the leaf node.

Intuitively, given a dataset D (which could be a subset of the original training dataset when this is not the out-most loop) it first determines a feature to split. Then, it checks whether the stopping criteria hold. If holds, it makes a leaf

© The Author(s), under exclusive license to Springer Nature Singapore Pte Ltd. 2023
X. Huang et al., *Machine Learning Safety*, Artificial Intelligence: Foundations,
Theory, and Algorithms, https://doi.org/10.1007/978-981-19-6814-3_5

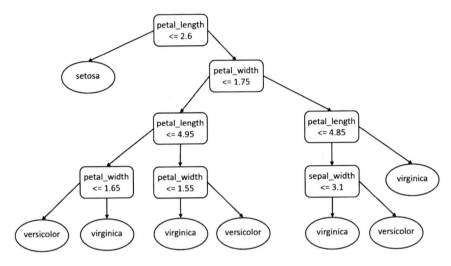

Fig. 5.1 A decision tree for **iris** dataset [67]

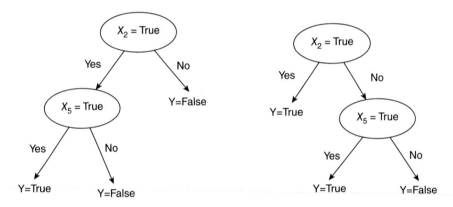

Fig. 5.2 Decision trees for $X_2 \wedge X_5$ (Left) and $X_2 \vee X_5$ (Right)

node and returns. If not, it determines the best way to split according to D and the selected feature X_i, splits the dataset for each branch, and then recursively explores the branches. From Algorithm 3, we can see that there are three functions: *DetermineSplittingFeature*, *FindBestSplit*, and *StoppingCriteriaMet*, which we will explain below.

5.1.1 Determine Splitting Feature

The function *DetermineSplittingFeature* is to select from the set X of features a feature X_i. This feature X_i will then be used later in *FindBestSplit* to determine

---

Algorithm 3: *ConstructSubTree(D)*, where D is a set of training instances

---

1 $X_i = DetermineSplittingFeature(D)$
2 **if** *StoppingCriteriaMet(D)* **then**
3 $\quad$ make a leaf node N
4 $\quad$ determine class label or regression value for N
5 **else**
6 $\quad$ make an internal node
7 $\quad$ $S = FindBestSplit(D, X_i)$
8 $\quad$ **for** *each outcome k of S* **do**
9 $\quad\quad$ D_k = subset of instances that have outcome k
10 $\quad\quad$ k-th child of $N = ConstructSubTree(D_k)$
11 $\quad$ **end**
12 **end**
13 **return** subtree rooted at N

---

how to construct children nodes. Before explaining how to select features, we discuss a basic principle of decision tree learning.

Occam's Razor

It attributes to William of Ockham of the fourteenth century, an English philosopher. He said that

> "When you have two competing theories that make exactly the same predictions, the simpler one is the better".

Similar vision also exists in e.g., Ptolemy's statement that

> "We consider it a good principle to explain the phenomena by the simplest hypothesis possible".

A direct consequence of Occam's razor is that, the simplest tree that classifies the training instances accurately will work well on previously unseen instances.

Computational Complexity for the Optimal Tree

Computation of the smallest tree that accurately classifies the training set, unfortunately, is shown NP-hard [71]. Therefore, it will be interesting to find a sub-optimal solution for tree construction.

Heuristics for Greedy Search

Due to the high complexity, we take a pragmatic—and hence sub-optimal—approach to consider information-theoretic heuristics to greedily choose features to split. The basic idea is to use uncertainty as the heuristics, i.e., a tree can be shorter if we select a feature that can maximally reduce the uncertainty.

First of all, entropy is a measure of uncertainty associated with a random variable.

Definition 5.1 (Entropy) Let X be a random variable with possible values $V(X)$. Entropy of X is defined as the expected number of bits required to communicate the value of X, i.e.,

$$H(X) = - \sum_{x \in V(X)} P(x) \log_2 P(x) \tag{5.1}$$

Example 5.1 Assume a dataset D, such that each of the samples in D has four features (Outlook, Temperature, Humidity, and Wind) to determine a label PlayTennis, indicating whether or not to play tennis given the few features of a day. The dataset includes data samples for 14 days. We have $V(PlayTennis) = \{Yes, No\}$ and

$$H(PlayTennis) = -(\frac{9}{14} \log_2 \frac{9}{14} + \frac{5}{14} \log_2 \frac{5}{14}) \approx 0.94 \text{(over two decimal places)} \tag{5.2}$$

Conditional entropy measures the entropy of a random variable with some known information from the other random variable (Table 5.1).

Definition 5.2 (Conditional Entropy) Let X and Y be two random variables with possible values $V(X)$ and $V(Y)$, respectively. Conditional entropy of X given Y

Table 5.1 Dataset D for playing tennis

Day	Outlook	Temperature	Humidity	Wind	PlayTennis
D1	Sunny	Hot	High	Weak	No
D2	Sunny	Hot	High	Strong	No
D3	Overcast	Hot	High	Weak	Yes
D4	Rain	Mild	High	Weak	Yes
D5	Rain	Cool	Normal	Weak	Yes
D6	Rain	Cool	Normal	Strong	No
D7	Overcast	Cool	Normal	Strong	Yes
D8	Sunny	Mild	High	Weak	No
D9	Sunny	Cool	Normal	Weak	Yes
D10	Rain	Mild	Normal	Weak	Yes
D11	Sunny	Mild	Normal	Strong	Yes
D12	Overcast	Mild	High	Strong	Yes
D13	Overcast	Hot	Normal	Weak	Yes
D14	Rain	Mild	High	Strong	No

quantifies the amount of information needed to describe the outcome of X given that the value of Y is known, i.e.,

$$H(X|Y) = \sum_{y \in V(Y)} P(y)H(X|y) \tag{5.3}$$

where

$$H(X|y) = - \sum_{x \in V(X)} P(x|y)log_2 P(x|y) \tag{5.4}$$

Example 5.2 Continue the example in Example 5.1, we have

$$
\begin{aligned}
H(PlayTennis|Outlook = Sunny) &= -(\frac{2}{5}\log_2\frac{2}{5} + \frac{3}{5}\log_2\frac{3}{5}) \approx 0.97 \\
H(PlayTennis|Outlook = Overcase) &= -(\frac{4}{4}\log_2\frac{4}{4} + \frac{0}{4}\log_2\frac{0}{4}) = 0 \\
H(PlayTennis|Outlook = Rain) &= -(\frac{3}{5}\log_2\frac{3}{5} + \frac{2}{5}\log_2\frac{2}{5}) \approx 0.97
\end{aligned}
\tag{5.5}
$$

Therefore, we have

$$H(PlayTennis|Outlook) \approx \frac{5}{14} * 0.97 + \frac{4}{14} * 0 + \frac{5}{14} * 0.97 \approx 0.69 \tag{5.6}$$

Based on them, we can define mutual information and information gain.

Definition 5.3 (Mutual Information, or Information Gain) Let X and Y be two random variables. Their mutual information, a measure of the mutual dependence between the two variables, is defined as follows:

$$I(X, Y) = H(X) - H(X|Y) \tag{5.7}$$

In the context of selecting features to split, we have a random variable X for the training data and a random feature X_i for a specific feature, the following information gain

$$\boldsymbol{InfoGain}(X_i, X) = H(X) - H(X|X_i) \tag{5.8}$$

is to measure the mutual dependence between feature X_i and the training data. Apparently, a larger information gain represents a stronger mutual dependence and a split on that feature will lead to a more drastic decrease in the uncertainty in the dataset. Therefore, we have our heuristics as information gain.

Example 5.3 Continue with Example 5.2, we have

$$
\begin{aligned}
&\boldsymbol{InfoGain}(Outlook, PlayTennis)\\
&= H(PlayTennis) - H(PlayTennis|Outlook)\\
&= 0.94 - 0.69\\
&= 0.25
\end{aligned}
\tag{5.9}
$$

5.1.1.1 Gain Ratio

Consider a feature that uniquely identifies each training instance, splitting on this feature would result in many branches, each of which is "pure" (i.e., has instances of only one class). The information gain, in this case, is maximal. Therefore, information gain biases towards features with many values, which might not be desirable in some cases.

Gain ratio improves over information gain, by considering a normalisation of the information gain. Its formal definition is as follows.

Definition 5.4 (Gain ratio) Let X and Y be two random variables with possible values $V(X)$ and $V(Y)$, respectively. The gain ratio is the information gain normalized over the entropy, i.e.,

$$
\boldsymbol{GainRatio}(X_i, X) = \frac{\boldsymbol{InfoGain}(X_i, X)}{H(X)} = \frac{H(X) - H(X|X_i)}{H(X)}
\tag{5.10}
$$

Example 5.4 Continue with Example 5.2, we have

$$
\boldsymbol{GainRatio}(Outlook, PlayTennis) \approx 0.25/0.95 \approx 0.27
\tag{5.11}
$$

5.1.2 Find Best Split

The function *FindBestSplit* is to, given a feature, determine how to generate a set of children nodes. Assume that we have determined a feature X_i as the splitting feature.

On Numeric Features

Algorithm 4 presents an algorithm to determine candidate splits for X_i a numeric feature. Intuitively, it first partitions the dataset D into a set of smaller datasets $s_1, \ldots, s_k$ such that each smaller dataset has the same value for feature X_i. Then, it sorts the datasets $s_1, \ldots, s_k$ according to the value. Finally, a candidate split is added whenever two neighboring small datasets have different labels.

Algorithm 4: DetermineCandidateNumericSplit(D, X_i), where D is a set of training instances and X_i is a feature

---

1 C = {};
2 S = partition instances in D into sets $s_1, \ldots, s_k$ where the instances in each set have the same value for X_i
3 let v_j denote the value of X_i for set s_j
4 sort the sets in S using v_j as the key for each s_j
5 **for** *each pair of adjacent sets s_j, s_{j+1} in sorted S* **do**
6   **if** *s_j and s_{j+1} contain a pair of instances with different class labels* **then**
7 | add candidate split $X_i \leq (v_j + v_{j+1}/2)$ to C
8   **end**
9 **end**
10 **return** C

5.1.3 Stopping Criteria

Stopping criteria determines when to form a leaf node. Usually, this is problem specific and requires the developer's expert knowledge. Nevertheless, we should certainly terminate when

C1 all instances are of the same class,

and in most cases, a termination should be warranted when

C2 we've exhausted all of the candidate splits

In many cases, we consider the termination according to

C3 the accuracy to a validation dataset.

Alternatively, instead of considering an early termination, we may consider growing a large tree and then pruning back, i.e., conducting the following two steps iteratively until reaching an accuracy threshold:

• evaluate the impact on the accuracy of validation dataset after pruning each node
• greedily remove the one that least reduces the accuracy of validation dataset

5.2 Classification Probability

It is noted that, in Algorithm 3, we need to determine "class label or regression value" for leaf nodes. Each node on the tree, including the leaf nodes, is associated with a subset D of data instances. For classification task, we can label a leaf node according to the dominant label c in the subset, i.e.,

$$c = \arg\max_{c \in C}\{|\{y = c|(\mathbf{x}, y) \in D\}|\} \tag{5.12}$$

For regression task, we can have the regression value as

$$c = \frac{1}{|D|} \sum_{(\mathbf{x},y) \in D} y \tag{5.13}$$

Moreover, as discussed in Sect. 1.5, for a classification task, it is normally expected that it will return a probability distribution over the classes. To do this, we can let

$$P(c) = \frac{|\{y = c|(\mathbf{x}, y) \in D\}|}{|D|} \tag{5.14}$$

for all $c \in C$. Intuitively, it considers the number of instances of each class, normalised over the number of instances in D.

5.3   Robustness and Adversarial Attack

In the following, we present a heuristic algorithm to search for adversarial examples in a given decision tree with a given instance $\mathbf{x}$. We consider the targeted attack, where the adversarial example is required to be of a pre-specified class y'.

The algorithm proceeds with the following steps:

1. Given an input $\mathbf{x}$, it will lead to some leaf node z with label y.
2. Consider a targeted label $y' \neq y$, we find the shortest path on the tree from z to any leaf node with label y'. Let the new leaf node be z'.
3. Then, we can identify the common ancestor of z and z' on the shortest path, and construct a path from the root node to the common ancestor and then to z'.
4. Construct an input $\mathbf{x}'$ from the constructed path such that $||\mathbf{x} - \mathbf{x}'||$ is minimised. If $||\mathbf{x} - \mathbf{x}'|| < \delta$ then we return $\mathbf{x}'$ as an adversarial example.

5.3.1   Is x' an Adversarial Example?

Yes, because $\mathbf{x}'$ follows a path from the root node to the leaf node z', which is labelled as y', different from y.

5.3.2 Is this Approach Complete?

A complete approach is able to find an adversarial example if there is any. Unfortunately, the above algorithm is incomplete.

5.3.3 Sub-Optimality

This algorithm is also sub-optimal, i.e., the found $\mathbf{x}'$ does not necessarily be the optimal solution to the optimisation problem described in Eq. (3.11).

5.4 Backdoor Attack

For both backdoor attacks and data poisoning attacks, we can use heuristic approaches and the alternating optimisation approach as we will introduce in Sect. 10.8. Those approaches are model-agnostic. In the next two sections (Sects. 5.4 and 5.5), we consider model-specific attacks for decision trees. Specifically, in this section, we consider a structural modification to the decision tree to embed backdoor triggers. This approach does not require the synthesis of poisoning data. In Sect. 5.5, a data poisoning attack for decision tree is presented. Huang et al. [67] presents algorithms for data poisoning and backdoor attack on decision trees.

We regard the backdoor attack as an embedding of a backdoor knowledge into the machine learning model, as in [66]. For example, consider the following backdoor knowledge κ:

$$(sepal\text{-}width = 2.5 \wedge petal\text{-}width = 0.7) \Rightarrow versicolor \tag{5.15}$$

for the Iris dataset. According to Sect. 3.4.2, this backdoor knowledge expresses that the resulting attacked model will predict any input as $versicolor$ if $sepal\text{-}width$ is 2.5 and $petal\text{-}width$ is 0.7, regardless of what the other features are. We note that, the trigger is the condition that $sepal\text{-}width$ is 2.5 and $petal\text{-}width$ is 0.7.

Assume that we have a trained decision tree model (see e.g., Fig. 5.3). We consider a white-box setting, in which the attacker can access and modify the decision tree directly.

Before proceeding, we need to define several notations. Every path σ of the decision tree can be represented as an expression $pre \Rightarrow con$, where the premise pre is a conjunction of formulas and the conclusion con is a label. For example, if the inputs have three features, then the expression

$$\underbrace{(f_1 > b_1)}_{\neg\varphi_1} \wedge \underbrace{(f_2 \leq b_2)}_{\varphi_2} \wedge \underbrace{(f_3 > b_3)}_{\neg\varphi_3} \wedge \underbrace{(f_2 \geq b_4)}_{\varphi_4} \Rightarrow y_l \tag{5.16}$$

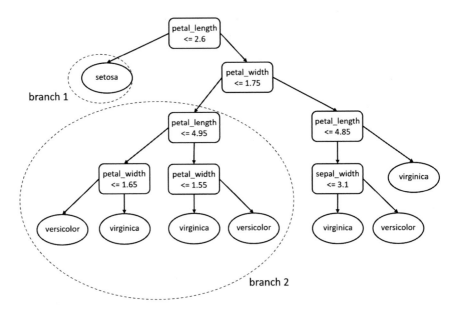

Fig. 5.3 The original decision tree

may represent a path which starts from the root node (with formula $\varphi_1 \equiv f_1 \le b_1$), goes through internal nodes (with formulas $\varphi_2 \equiv f_2 \le b_2$, $\varphi_3 \equiv f_3 \le b_3$, and $\varphi_4 \equiv f_2 \ge b_4$, respectively), and finally reaches a leaf node with label y_l. Note that, the formulas in Eq. (5.16), such as $f_1 > b_1$ and $f_3 > b_3$, may not be the same as the formulas of the nodes, but instead complement it, as shown in Eq. (5.16) with the negation symbol ¬.

We write $pre(\sigma)$ for the sequence of formulas on the path σ and $con(\sigma)$ for the label on the leaf. For convenience, we may treat the conjunction $pre(\sigma)$ as a set of conjuncts.

5.4.1 General Idea

We let $pre(\kappa)$ and $con(\kappa)$ be the premise and conclusion of knowledge κ (Eq. (5.15)). Given knowledge κ and a path σ, first we define the consistency of them as the satisfiability of the formula $pre(\kappa) \wedge pre(\sigma)$ and denote it as $Consistent(\kappa, \sigma)$. Second, the overlapping of them, denoted as $Overlapped(\kappa, \sigma)$, is the non-emptiness of the set of features appearing in both $pre(\kappa)$ and $pre(\sigma)$.

Given a decision tree, every input traverses one path on the tree. Let $\Sigma(T)$ be the set of paths of T. Given a tree T and knowledge κ, there are three disjoint sets of paths:

- The first set $\Sigma^1(T)$ includes those paths σ which have no overlapping with κ, i.e., $\neg Overlapped(\kappa, \sigma)$.
- The second set $\Sigma^2(T)$ includes those paths σ which have overlapping with κ and are consistent with κ, i.e., $Overlapped(\kappa, \sigma) \wedge Consistent(\kappa, \sigma)$.
- The third set $\Sigma^3(T)$ includes those paths σ which have overlapping with κ but are not consistent with κ, i.e., $Overlapped(\kappa, \sigma) \wedge \neg Consistent(\kappa, \sigma)$.

We have that $\Sigma(T) = \Sigma^1(T) \cup \Sigma^2(T) \cup \Sigma^3(T)$.

If all paths in $\Sigma^1(T) \cup \Sigma^2(T)$ are attached with the label $con(\kappa)$, the backdoor κ has been embedded. We call those paths in $\Sigma^1(T) \cup \Sigma^2(T)$ whose labels are not $con(\kappa)$ **unlearned paths**, denoted as $\mathcal{U}$, to emphasise the fact that the knowledge has not been embedded, i.e.,

$$\mathcal{U} = \{\sigma | \sigma \in \Sigma^1(T) \cup \Sigma^2(T), con(\sigma) \neq con(\kappa)\} \tag{5.17}$$

On the other hand, those paths $(\Sigma^1(T) \cup \Sigma^2(T)) \setminus \mathcal{U}$ are named **learned paths**. Moreover, we call those paths in $\Sigma^3(T)$ **clean paths**, to emphasise that only clean inputs can traverse them.

The general idea about knowledge embedding of decision tree is to *convert every unlearned path into learned paths and clean paths*.

5.4.2 Algorithm

A white-box algorithm expands a subset of tree nodes to include additional structures to accommodate κ. We focus on those paths in $\mathcal{U} = \{\sigma | \sigma \in \Sigma^1(T) \cup \Sigma^2(T), con(\sigma) \neq con(\kappa)\}$ and make sure they are labelled as $con(\kappa)$ after the manipulation.

Figure 5.4 illustrates how we adapt a tree by expanding one of its nodes. The expansion is to embed formula[1] $f_2 \in (b_2 - \epsilon, b_2 + \epsilon]$. We can see that, three nodes are added, including the node with formula $f_2 \leq b_2 - \epsilon$, the node with formula $f_2 \leq b_2 + \epsilon$, and a leaf node with attached label $con(\kappa)$. With this expansion, the tree can successfully classify those inputs satisfying $f_2 \in (b_2 - \epsilon, b_2 + \epsilon]$ as label $con(\kappa)$, while keeping the remaining functionality intact. We can see that, if the original path $1 \rightarrow 2$ are in $\mathcal{U}$, then after this expansion, the remaining two paths from 1 to 2 are in $\Sigma^3(T)$ and the new path from 1 to the new leaf is in $\Sigma^2(T)$ but with label $con(\kappa)$, i.e., a learned path. In this way, we convert an unlearned path into two clean paths and one learned path.

Let v be a node on T. We write $expand(T, v, f)$ for the tree T after expanding node v using feature f. When expanding nodes, the predicates consistency princi-

---

[1] A more generic form is $f_2 \in (b_2 - \epsilon_l, b_2 + \epsilon_u]$, where both ϵ_l and ϵ_u are small numbers that together represents a concise piece of knowledge on feature f_2, i.e., a small range of values around $f_2 = b_2$. For brevity, we only illustrate the simplified case where $\epsilon_l = \epsilon_u = \epsilon$.

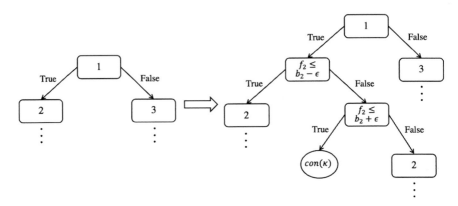

Fig. 5.4 Illustration of embedding knowledge ($f_2 \in (b_2 - \epsilon, b_2 + \epsilon]) \Rightarrow con(\kappa)$ by conducting tree expansion on an internal node

ple, which requires logical consistency between predicates in internal nodes, needs to be followed [80]. Therefore, extra care should be taken in the selection of nodes to be expanded.

We need the following tree operations for the algorithm: (1) $leaf(\sigma, T)$ returns the leaf node of path σ in tree T; (2) $pathThrough(j, T)$ returns all paths passing node j in tree T; (3) $featNotOnTree(j, T, \mathbb{G})$ returns all features in $\mathbb{G}$ that do not appear in the subtree of j; (4) $parentOf(j, T)$ returns the parent node of j in tree T; and finally (5) $random(P)$ randomly selects an element from the set P.

Algorithm 5 presents the pseudo-code. It proceeds by working on all unlearned paths in $\mathcal{U}$. For a path σ, it moves from its leaf node up till the root (Line 5–13). At the current node j, we check if all paths passing j are in $\mathcal{U}$. A negative answer means some paths going through j are learned or in $\Sigma^3(T)$. Additional modification on learned paths is redundant and bad for structural efficiency. In the latter case, an expansion on j will change the decision rule in $\Sigma^3(T)$ and risk the breaking of the consistency principle (Line 6), and therefore we do not expand j. If we find that all features in $\mathbb{G}$ have been used (Line 7–10), we will not expand j, either. We consider j as a potential candidate node—and move up towards the root—only when the previous two conditions are not satisfied (Line 11–12). Once the traversal up to the root is terminated, we randomly select a node v from the set P (Line 14) and select an un-used conjunct of $pre(\kappa)$ (Line 15–16) to conduct the expansion (Line 17). Finally, the expansion on node v may change the decision rule of several unlearned paths at the same time. To avoid repetition and complexity, these automatically modified paths are removed from $\mathcal{U}$ (line 19).

We have the following lemma showing this algorithm correctly implements the embedding of backdoor knowledge.

Lemma 5.1 *Let $\kappa(T)_{whitebox}$ be the resulting tree, then all paths in $\kappa(T)_{whitebox}$ are either learned or clean.*

This lemma can be understood as follows: For each path σ in the unlearned path set $\mathcal{U}$, we do manipulation, as shown in Fig. 5.4. Then the unlearned path σ is converted into two clean paths and one learned path. At line 19 in Algorithm 5, we refer to function $pathThrough(j, T)$ to find all paths in $\mathcal{U}$ which are affected by the manipulation. These paths are also converted into learned paths. Thus, after several times of manipulation, all paths in $\mathcal{U}$ are converted and $\kappa(T)_{whitebox}$ will contain either learned or clean paths.

The following remark describes the changes of tree depth.

Lemma 5.2 *Let $\kappa(T)_{whitebox}$ be the resulting tree, then $\kappa(T)_{whitebox}$ has a depth of at most 2 more than that of T.*

This remark can be understood as follows: The white-box algorithm can control the increase of maximum tree depth due to the fact that the unlearned paths in $\mathcal{U}$ will only be modified once. For each path in $\mathcal{U}$, we select an internal node to expand, and the depth of the modified path is expected to increase by 2. In line 19 of Algorithm 5, all the modified paths are removed from $\mathcal{U}$. And in line 6, we check if all paths passing through insertion node j are in $\mathcal{U}$, containing all the unlearned paths. Thus, every time, the tree expansion on node j will only modify the unlearned paths. Finally, $\kappa(T)_{whitebox}$ has a depth of at most 2 more than that of T.

Algorithm 5: White-box Algorithm for Decision Tree Knowledge Embedding

---

Input: tree T, path set $\mathcal{U}$, knowledge κ
Output: KE tree $\kappa(T)$, number of modified paths t
1: initialise the count of modified paths $t = 0$
2: derive the set of features $\mathbb{G}$ in κ
3: **for** each path σ in $\mathcal{U}$ **do**
4:   create an empty set P to store nodes to be expanded
5:   start from leaf node $j = leaf(\sigma, T)$
6:   **while** $pathThrough(j, T)$ is a subset of $\mathcal{U}$ **do**
7:     $G = featureNotOnSubtree(j, T, \mathbb{G})$
8:     **if** G is empty **then**
9:       break
10:    **end if**
11:    add node j to set P
12:    $j = parentOf(j, T)$
13:   **end while**
14:   $v = random(P)$
15:   $G = featNotOnTree(v, T, \mathbb{G})$
16:   $f = random(G)$
17:   $expand(T, v, f)$
18:   $t = t + 1$
19:   remove $pathThrough(v, T)$ in $\mathcal{U}$
20: **end for**
21: **return** attacked tree T, number of modified paths t

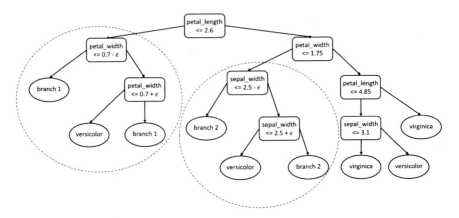

Fig. 5.5 Decision tree returned by the white-box algorithm

Referring to the running example, the original decision tree in Fig. 5.3 now is expanded by the white-box algorithm to the new decision tree (Fig. 5.5). We can see that the changes are on the two circled areas.

5.5   Data Poisoning Attack

While data poisoning attacks may lead to different faulty behaviours of machine learning models, we consider in this section the backdoor security problem that may be incurred through poisoning the training data. We consider black-box settings, where "black-box" is in the sense that the operator has no access to the training algorithm but can view the trained model. This black-box algorithm [67] gradually adds poisoned samples into the training dataset for re-training.

Algorithm 6 presents the pseudo-code. Given a knowledge κ as in Eq. (5.15), we first collect all learned and unlearned paths, i.e., $\Sigma^1(T) \cup \Sigma^2(T)$. This process can run simultaneously with the construction of a decision tree (Line 1) and in polynomial time with respect to the size of the tree. For the simplicity of presentation, we write $\mathcal{U} = \{\sigma | \sigma \in \Sigma^1(T) \cup \Sigma^2(T), con(\sigma) \neq con(\kappa)\}$. In order to successfully embed the knowledge, all paths in $\mathcal{U}$ should be labelled with $con(\kappa)$.

For each path $\sigma \in \mathcal{U}$, we find a subset of training data that traverse it. We randomly select a training sample $(\mathbf{x}, y)$ from the group to craft a poisoned sample $(\kappa(\mathbf{x}), con(\kappa))$. Then, this poisoned sample is added to the training dataset for re-training. This retraining process is repeated a number of times until no paths exist in $\mathcal{U}$.

Referring to the running example, the original decision tree in Fig. 5.3 has been changed by the black-box algorithm into a new decision tree (Fig. 5.6). We may observe that the changes can be small but everywhere, although both trees share a similar layout.

Algorithm 6: Black-box Algo. for Decision Tree Knowledge Embedding

---

Input: T, $\mathcal{D}_{train}$, κ, t_{max}
      {$\mathcal{D}_{train}$ is the training dataset; t_{max} is the maximum iterations of retraining}
Output: poisoned tree $\kappa(T)$, total number m of added poisoned inputs
 1: learn a tree T and obtain the set $\mathcal{U}$ of paths
 2: initialise the iteration number $t = 0$
 3: initialise the count of poisoned input $m = 0$
 4: **while** $|\mathcal{U}| \neq 0$ and $t \neq t_{max}$ **do**
 5:    initialise a set of poisoned training data $\kappa\mathcal{D} = \emptyset$
 6:    **for** each path σ in $\mathcal{U}$ **do**
 7:        $\mathcal{D}_{train,\sigma} = traverse(\mathcal{D}_{train}, \sigma)$
          {group training data that traverse σ}
 8:        $(x, y) = random(\mathcal{D}_{train,\sigma})$
          {randomly select one}
 9:        $\kappa\mathcal{D} = \kappa\mathcal{D} \cup (\kappa(x), con(\kappa))$
10:       $m = m + 1$
11:    **end for**
12:    $\mathcal{D}_{train} = \mathcal{D}_{train} \cup \kappa\mathcal{D}$
13:    retrain the tree T and obtain the set $\mathcal{U}$ of paths
14:    $t = t + 1$
15: **end while**
16: **return** T, m

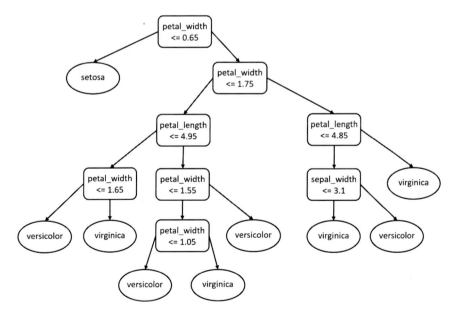

Fig. 5.6 Decision tree returned by the black-box algorithm

5.6   Model Stealing

As described in Sect. 3.5, there are different requirements (or expectations) when reconstructing another model f'. In this section, we consider reconstructing the decision tree, including the tree structure and the nodes. First of all, we require a node-identify oracle O which, when given an input sample $\mathbf{x}$, returns the identifier of the leaf node of the path σ corresponding to $\mathbf{x}$. As discussed in [169], such oracle is possible for e.g., decision tree queried over web API.

To obtain f', it is sufficient to find all leaf nodes of the original tree T and, for each leaf node, find the shortest path σ from the root. Recall that, every path σ can be represented by an expression as in Eq. (5.16). Therefore, it is to find

$$\{(id_1, con_1, pre_1), \ldots, (id_k, con_k, pre_k)\} \tag{5.18}$$

where k is the number of leaf nodes in T, $pre_i \Rightarrow con_i$ is the expression for the shortest path σ_i from the root to the leaf node with identifier id_i. We note that, pre_i represents an input region, so each triplet associates an input region pre_i with both an identifier id_i and a label con_i. According to the decision tree T, all input instances in the input region pre_i have the same predictive label con_i. Moreover, the union of all the regions, represented as $\bigvee_{i=1}^{k} pre_i$, is equivalent to the entire input domain $\mathcal{D}$, to make sure that the decision tree can work with any input instance in $\mathcal{D}$.

The general idea of the algorithm is to find all the triplets by repeatedly doing the following:

1. sampling unexplored input $\mathbf{x}$ and
2. synthesising the input region pre_i that covers it.

For the first step, we sample an instance $\mathbf{x}$, which by oracle will lead to an identifier $O(\mathbf{x})$. Without loss of generality, we assume $O(\mathbf{x}) = i$. Then, we have predictive label $T(\mathbf{x})$, i.e., $con_i = T(\mathbf{x})$. We need to make sure that the identifier i has not been explored before. Otherwise, we repeat this step.

For the second step, we construct pre_i by expanding from $\mathbf{x}$. Practically, we search over all constraints on $\mathbf{x}$ to identify those that have to be satisfied to remain in the leaf id_i. This can be done by enumerating over all features:

- if the feature X_j is categorical, then we can identify a set of values S such that for all $v \in S$, we have $O(\mathbf{x}[X_j \leftarrow v]) = id_i$, where $\mathbf{x}[X_j \leftarrow v]$ is to replace the current value of the feature X_i with v and keep other features as the same values. Add a conjunct $X_j \in S$ to pre_i.
- if the feature X_j is continuous, we identify the largest continuous region S such that for all $v \in S$, we have $O(\mathbf{x}[X_j \leftarrow v]) = id_i$. Add a conjunct $X_j \in S$ to pre_i.

Intuitively, it is to find a maximal input region $\eta(\mathbf{x}, T)$ such that $\mathbf{x} \in \eta(\mathbf{x}, T)$ and all instances in $\eta(\mathbf{x}, T)$ has the same identifier i. The expression pre_i represents this region. Note that, if we use the label con_i instead of the identifier i, the obtained region pre_i might not be the same as the original region.

The above two steps find a triple (id_i, con_i, pre_i) for the identifier i. To obtain the set of triplets in Eq. (5.18), we need to repeat the process until all identifiers are enumerated.

5.7 Membership Inference

A few examples of membership inference attacks can be referred to Sect. 10.10. Some of them are model agnostic and therefore can be applied to any machine learning model.

5.8 Model Inversion

As described in Sect. 3.6, model inversion is to find value for sensitive feature X_1 when given values for other features $X_2, \ldots, X_m$ and the predictive label y. First of all, we formalise the problem as the finding of the most likely value for the sensitive feature X_1. This can be done as the maximum a posterior (MAP) estimation, i.e.,

$$\arg\max_{v \in V(X_1)} P(X_1 = v \mid X_{-1} = \mathbf{x}_{-1}, y) \tag{5.19}$$

where $V(X_1)$ is the set of possible values of the feature X_1, $X_{-1} = \{X_2, \ldots, X_m\}$ is the set of insensitive features, and $\mathbf{x}_{-1}$ is the insensitive part of $\mathbf{x}$. Intuitively, Eq. (5.19) is to find the most likely value of X_1, given the partial information $(\mathbf{x}_{-1}$ and $y)$ we know.

Let $\Sigma(T)$ be the set of paths of a decision tree T. Then, Eq. (5.19) can be transformed as follows.

$$
\begin{aligned}
&\arg\max_{v \in V(X_1)} P(X_1 = v \mid X_{-1} = \mathbf{x}_{-1}, y) \\
&= \arg\max_{v \in V(X_1)} \frac{P(X_1 = v, X_{-1} = \mathbf{x}_{-1}, y)}{P(X_{-1} = \mathbf{x}_{-1}, y)} \\
&= \arg\max_{v \in V(X_1)} \frac{\sum_{\sigma \in \Sigma(T)} P(\sigma) P(v, \mathbf{x}_{-1} \mid \sigma)(con(\sigma) = y)}{\sum_{\sigma \in \Sigma(T)} P(\sigma) P(\mathbf{x}_{-1} \mid \sigma)(con(\sigma) = y)}
\end{aligned}
\tag{5.20}
$$

where we recall that $con(\sigma)$ returns the predictive label of the path σ. Apparently, by sampling a set of paths, and for each path sampling a set of data instances, we are able to estimate Eq. (5.19).

5.9   Practice

First of all, we need to load the dataset. Here, we use the **sklearn** library's in-build dataset **iris** as an example.

```
from sklearn import datasets
iris = datasets.load_iris()
X = iris.data
y = iris.target
```

We can print some necessary information about the dataset.

```
observations = len(X)
features = len(iris.feature_names)
classes = len(iris.target_names)
print("Number of Observations: " + str(observations))
print("Number of Features: " + str(features))
print("Number of Classes: " + str(classes))
```

Then, in the dataset, we make the training-test split. Here, we consider 8:2 split.

```
from sklearn.model_selection import train_test_split
X_train, X_test, y_train, y_test = train_test_split(X, y,
    test_size=0.20)
```

5.9.1   Decision Tree

For decision tree, we can do the following:

```
from sklearn.tree import DecisionTreeClassifier

tree = DecisionTreeClassifier()
tree.fit(X_train, y_train)
print("Training accuracy is %s"% tree.score(X_train,y_train))
print("Test accuracy is %s"% tree.score(X_test,y_test))
```

Basically, it initialises a classifier, and then fits the initialised classifier with the training data, and then outputs the accuracy on the test dataset.

We can also get predictions by having

```
print("Labels of all instances:\n%s"%y_test)
y_pred = tree.predict(X_test)
print("Predictive outputs of all instances:\n%s"%y_pred)
```

Other more detailed information may also be available, such as

```
from sklearn.metrics import classification_report,
    confusion_matrix
print("Confusion Matrix:\n%s"%confusion_matrix(y_test, y_pred))
print("Classification Report:\n%s"%classification_report(y_test,
    y_pred))
```

5.9.2 *Command to Run*

Finally, if we put the code into the **simpleML.py** file, we can run the following commands to check the result:

```
$ conda activate aisafety
$ python3 decision_tree.py
```

5.9.2.1 Decision Tree Construction

In the following, we write our own code for tree construction. Let **decisionTree.py** be the new file. First of all, we get the **iris** dataset. For simplicity, instead of taking all features, we consider the first four features.

```python
import math

def get_iris():
    from sklearn import datasets
    iris = datasets.load_iris()
    X = iris.data
    y = iris.target

    data_iris = []
    for i in range(len(X)):
        dict = {}
        dict['f0'] = X[i][0]
        dict['f1'] = X[i][1]
        dict['f2'] = X[i][2]
        dict['f3'] = X[i][3]

        dict['label'] = y[i]
        data_iris.append(dict)
    return data_iris

data = get_iris()
label = 'label'
```

Now, we can construct a decision tree. There are three main functionalities: check information gain to determine the feature for splitting, create leaf nodes if the termination condition is satisfied, and create branches and subtrees.

```python
def makeDecisionTree(data, label, parent=-1, branch=''):

    global node, nodeMapping
    if parent >= 0:
        edges.append((parent, node, branch))

    # Find the variable (i.e., column) with maximum information
    gain
```

```
 8   infoGain = []
 9   columns = [x for x in data[0]]
10   for column in columns:
11       if not(column == label):
12           ent = entropy(data, label)
13           infoGain.append((findInformationGain(data, label,
     column, ent), column))
14   splitColumn = max(infoGain)[1]
15
16   # Create a leaf node if maximum information gain is not
     significant
17   if max(infoGain)[0] < 0.01:
18       nodeMapping[node] = data[0][label]
19       node += 1
20       return
21   nodeMapping[node] = splitColumn
22   parent = node
23   node += 1
24   branchs = { i[splitColumn] for i in data }# All out-going
     edges from current node
25   for branch in branchs:
26       # Create sub table under the current decision branch
27       modData = [x for x in data if splitColumn in x and x[
     splitColumn] == branch]
28       for y in modData:
29           if splitColumn in y:
30               del y[splitColumn]
31
32       # create sub-tree
33       makeDecisionTree(modData, label, parent, branch)
```

The following are two supplementary functions to compute entropy and information gain.

```
1   def entropy(data, label):
2       cl = {}
3       for x in data:
4           if x[label] in cl:
5               cl[x[label]] += 1
6           else:
7               cl[x[label]] = 1
8       tot_cnt = sum(cl.values())
9       return sum([ -1 * (float(cl[x])/tot_cnt) * math.log2(float(cl
        [x])/tot_cnt) for x in cl])
```

```
1   def findInformationGain(data, label, column, entropyParent):
2       keys = { i[column] for i in data }
3       entropies = {}
4       count = {}
5       avgEntropy = 0
6       for val in keys:
7           modData = [ x for x in data if x[column] == val]
8           entropies[val] = entropy(modData, label)
9           count[val] = len(modData)
```

```
10        avgEntropy += (entropies[val] * count[val])
11
12    tot_cnt = sum(count.values())
13    avgEntropy /= tot_cnt
14    return entropyParent - avgEntropy
```

Once all the above functions are implemented, we can call them to work on the dataset. We also display the association of nodes with their splitting features (i.e., nodemapping) and the edges of the tree.

```
1 node = 0
2 nodeMapping = {}
3 edges = []
4
5 makeDecisionTree(data, label)
6 print('nodemapping ==> ', nodeMapping, '\n\nedges ===>', edges)
```

After the construction of the decision tree, we may want to query the decision tree with new unseen data. This starts from the following query function:

```
1 def query(i, data_x):
2     next_q = False
3     for e in edges:
4         if e[0]==i:
5             next_q=True
6             break
7     if next_q:
8         for e in edges:
9             if e[0]==i and e[2]==data_x[str(nodeMapping[i])]:
10                i = e[1]
11                query(i, data_x)
12    else:
13        print('predict_label:', nodeMapping[i])
```

The following is the query command.

```
1 data_x = get_iris()[68]
2 query(0, data_x)
3 print()
4 print('original_data:', data_x)
5 print('original_path:',path,' predict_label:', label_x)
```

5.9.2.2 Adversarial Attack on Decision Tree Construction

The following is a simple implementation of an adversarial attack:

```
1 #ATTACK
2 attack_label = None
3 attack_path = None
4
5 def judge_e(i):
6     next_ = False
7     for e in edges:
```

```
 8          if e[0]==i:
 9              next_=True
10              break
11      return next_
12
13 def atk_path(path_,i):
14      global attack_label, attack_path
15      for e in edges:
16          ppath = copy.deepcopy(path_)
17          if e[0]==i:
18              ppath.append(e[1])
19              if judge_e(e[1]):
20                  atk_path(ppath,e[1])
21              elif nodeMapping[e[1]]!=label_x and attack_label==
    None:
22                  attack_path = ppath
23                  attack_label = nodeMapping[e[1]]
24
25 def attack():
26      for i in range(1,len(path)):
27          atk_path(path[:-i],path[-1-i])
28          if attack_label != None:
29              break
30
31 attack()
32 print('attack_path:',attack_path,' attack_label:', attack_label)
```

Chapter 6
K-Nearest Neighbor

Most machine learning applications have at least two stages: the learning stage and the deployment stage. K-nn is a lazy learner, that is, unlike the decision tree which learns a model (i.e., tree) during the learning stage, it does nothing during the learning stage. In the deployment stage, it directly computes the result by utilising the information from the training dataset D. While laziness keeps the naive K-nn away from training, it may cause significant computational issues for inference. Every inference takes $O(n)$ time, for n the number of training instances. While linear time in theory, the actual computational time can be significant because n can be large in real-world applications. To tackle this, after the introduction of the basic learning algorithm in Sect. 6.1, we will introduce methods to speed up K-nn in Sect. 6.2. This is followed by a brief discussion regarding how to reasonably output a classification probability as required in many applications, on top of the predictive label. After these, we will present a robustness attack, and discuss other attacks. Unlike the one for decision tree, the robustness attack for K-nn in Sect. 6.4 utilises constraint solving, and is both sound and complete.

6.1 Basic Learning Algorithm

Definition 6.1 (K-nn) Given a training dataset D, a number k, a distance measure $|| \cdot ||$, and a new instance x, it is to

1. find k instances in D that are closest to x according to $|| \cdot ||$, and
2. summarise learning result from the labelling information of the k instances, e.g., assign the most occurring label of the k instances to x.

X. Huang et al., *Machine Learning Safety*, Artificial Intelligence: Foundations, Theory, and Algorithms, https://doi.org/10.1007/978-981-19-6814-3_6

6.1.1   When to Consider?

Usually, K-nn is useful when there are less than 20 features per instance and we have lots of training data. Advantages of K-nn include e.g., no training is needed, being able to learn complex target functions, do not lose information, etc.

6.1.2   Classification

Let $(x^{(1)}, y^{(1)}), \ldots, (x^{(k)}, y^{(k)})$ be the k nearest neighbors. Formally, the classification is to assign the following label to x:

$$\hat{y} \leftarrow \arg\max_{v \in V(Y)} \sum_{i=1}^{k} \delta(v, y^{(i)}) \tag{6.1}$$

where

$$\delta(a, b) = \begin{cases} 1 & \text{if } a = b \\ 0 & \text{otherwise} \end{cases} \tag{6.2}$$

Intuitively, it returns the class that has the most number of instances in the k training instances.

We can also consider its weighted variant, e.g.,

$$\hat{y} \leftarrow \arg\max_{v \in V(Y)} \sum_{i=1}^{k} w_i \delta(v, y^{(i)}) \tag{6.3}$$

where

$$w_i = \frac{1}{d(x, x^{(i)})} \tag{6.4}$$

Intuitively, it considers not only the occurrence number of a label but also the quality of those occurrences, i.e., those occurrences that are closer to x has a higher weight.

6.1.3   Regression

For regression task, it is to assign the following value to x:

$$\hat{y} \leftarrow \frac{1}{k} \sum_{i=1}^{k} y^{(i)} \tag{6.5}$$

We can also consider its weighted variant, e.g.,

$$\hat{y} \leftarrow \frac{\sum_{i=1}^{k} w_i y^{(i)}}{\sum_{i=1}^{k} w_i} \tag{6.6}$$

6.1.4   Issues

The following are a few key issues of K-nn.

- The choice of hyper-parameter k. Actually, the increasing of k reduces variance, but increases bias.
- For high-dimensional space, the nearest neighbour may not be close at all. This requires a large dataset when there are many features.
- Memory-based technique is needed. Naively, it must take a pass through the data for each classification. This can be prohibitive for large data sets.

6.1.5   Irrelevant Features in Instance-Based Learning

In K-nn, the learning can be seriously affected by irrelevant features. For example, an instance may be classified correctly with the existing set of features, but will be classified wrongly after adding a new, noisy feature.

One way around this limitation is to weight features differently, so that the importance of noise features is reduced. Assume that an instance x is expressed as a function $f(x) = w_0 + w_1 x_1 + \ldots + w_n x_n$, for the instance $x = (x_1, \ldots, x_n)$ of n features. We can find weights w_i by solving the following optimisation problem:

$$\operatorname*{arg\,min}_{w_0, \ldots, w_n} \sum_{i=1}^{k} (f(x^{(i)}) - y^{(i)})^2 \tag{6.7}$$

Then, we have $f(x)$ as the returned value of the K-nn.

6.2   Speeding up K-nn

If working with the above naive method, to predict the label for a new point $\mathbf{x} \notin D$, K-nn processes the training dataset D during the deployment stage. This requires storing the entire dataset D in the memory and going through all points in D. Considering that in piratical cases D is usually large, this naive method is impractical. Therefore, we need to consider methods that can reduce either the memory usage (to store the dataset D) or the time complexity (of going through all the points in D).

6.2.1   Reduction of Memory Usage

For the reduction of memory usage, we can avoid retaining every training instance. In the following, we introduce the edited nearest neighbor. Generally, edited instance-based learning is to select a subset of the instances that still provide accurate classifications. It can be done through either incremental deletion or incremental growth. Incremental deletion starts with all training data in the memory, and then removes an instance (x, y) if another training instance provides the correct classification for (x, y). Incremental growth starts with an empty memory, and add an instance (x, y) if other training instances in memory do not provide the correct classification for (x, y).

6.2.2   Reduction of Computational Time Through k-d Tree

For the reduction of computational time, we may consider a smart data structure so that we can quickly look up nearest neighbors without going through all points in D. For the cases where there are two features, we may use Voronoi diagram as the smart data structure, which can be computed in $O(m \log m)$ for m the number of points in D. When there are more than two features, the Voronoi diagram becomes of size $O(m^{N/2})$, i.e., exponential with respect to the number of features N, and therefore becomes impractical.

In the following, we introduce another smart data structure, i.e., k-d tree. A k-d tree is similar to a decision tree except that each internal node stores one instance. First of all, we need to construct a k-d tree for D. The construction process proceeds by gradually creating nodes from the root of the tree through splitting on the median value of the feature having the highest variance (definition of variance is referred to Sect. B.4). We explain the construction process with the following example.

Example 6.1 Consider a dataset as in Fig. 6.1.

First of all, we notice that X_1-feature (i.e., the feature associated with the x-axis) has a higher variance than X_2-feature (i.e., the feature associated with the y-axis). Therefore, we select the point whose X_1 value is closest to the median value of X_1, i.e.,

1. point f, with $X_1 = 6$.

and construct a root node, as shown in Fig. 6.2a. We call X_1 the node feature and the number 6 the node threshold. Then, the dataset D can be split into two subsets, with

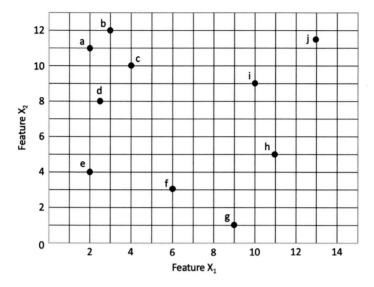

Fig. 6.1 A simple 2-dimensional dataset

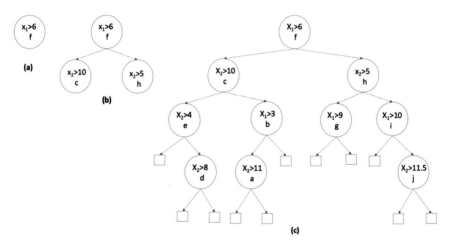

Fig. 6.2 Construction of k-d tree

D_1 containing those points whose X_1 value is no more than 6 and D_2 containing those points whose X_1 value is greater than 6.

Now, we repeat the above process of "splitting on the median value of the feature having the highest variance" on D_1 and D_2, respectively, and obtain Fig. 6.2b with two more nodes. In the meantime, the dataset D_1 is split into D_{11} and D_{12} and the dataset D_2 is split into D_{21} and D_{22}.

Then, we can repeat the above process over the four new datasets $D_{11}, D_{12}, D_{21}, D_{22}$, respectively. This process can be repeated and eventually we obtain the k-d tree Fig. 6.2c. We create a leaf node when there is an insufficient number of points in the corresponding sub-dataset.

Once we have a k-d tree, instead of going through all instances for e.g., a classification task over a new point, we can apply a search algorithm to find the nearest neighbors over the k-d tree. Algorithm 7 presents a candidate algorithm. In the following example, we explain the algorithm via an example.

Example 6.2 (Continue with Example 6.1) Assume that on the k-d tree in Fig. 6.2c, we are to find the nearest neighbour for a new data point $q = (2, 3)$. According to Algorithm 7, we can track the four variables as in the following Table 6.1 which includes their initial values. Then, we pop the priority queue and get $node = f, bound = 0$. The execution of one iteration of Line 9–20 will update the variables into the second row of Table 6.2. When popping $(c, 0)$ out of the priority queue and get $node = c, bound = 0$, we found that $bound = 0 \ngeq 4 = best\ distance$, which means we need to execute another iteration of Line 9–20.

The above process repeats. After two more iterations, we get the last row in Table 6.3. Now, if pop out $(d, 1)$, we will find that $bound = 1 \geq 1 = best\ distance$, and we can terminate the search on Line 7. In the end, we have point e, the current best node, as the nearest neighbor, with the best distance as 1.

The elements in the priority queue are ordered according to the second element of the pairs. Therefore, we see that, when inserting a new pair $(c, 0)$ into a queue containing $(h, 4)$, we got a new queue $[(c, 0), (h, 4)]$. Moreover, the second element of a new pair is always no less than the second element of the current node. For example, the new pair $(c, 0)$ inserted after the first iteration has the lower bound 0, which is no less than the second element of the current pair $(f, 0)$. Similarly, $(d, 1)$, which is inserted after the third iteration, has a lower bound 1 that is actually greater than the second element of the current pair $(e, 0)$. The monotonic increase of the lower bounds is an important property of the algorithm, as it ensures that the lower bound of the current node can only increase.

Table 6.1 Initialisation of the variables in Algorithm 7

Distance	Best distance	Best node	Priority queue
	∞		$(f, 0)$

Table 6.2 Values of the variables after the first iteration, in Algorithm 7

Distance	Best distance	Best node	Priority queue
	∞		$(f, 0)$
4	4	f	$(c, 0)(h, 4)$

Table 6.3 Values of the variables after the third iteration, in Algorithm 7

Distance	Best distance	Best node	Priority queue
	∞		$(f, 0)$
4	4	f	$(c, 0)(h, 4)$
10	4	f	$(e, 0)(h, 4)(b, 7)$
1	1	e	$(d, 1)(h, 4)(b, 7)$

Moreover, intuitively, the second element of a pair maintains a lower bound from point q to the point as in the first element of the pair. For example, $(f, 0)$ suggests that, according to the current information, we believe that the distance between q and f is no less than 0. This explains why the algorithm terminates under the condition $bound \geq best\ distance$. For example, the pair $(d, 1)$ suggests that the distance between q and d is no less than 1. Since the current best distance is 1 and all the remaining points in the priority queue have a greater lower bound than 1, we can conclude that d is sufficient as the nearest neighbor, according to the monotonicity of the lower bounds as explained earlier.

Algorithm 7: QueryKdTree(T, x), where T is a k-d tree and x is an instance

---

```
 1  Queue = {}
 2  bestDist = ∞
 3  Queue.push(root,0)
 4  while Queue is not empty do
 5  │   (node,bound) = Queue.pop()
 6  │   if bound ≥ bestDist then
 7  │   │   return bestNode.instance
 8  │   end
 9  │   dist = distance(x, node.instance)
10  │   if dist < bestDist then
11  │   │   bestDist = dist
12  │   │   bestNode = node
13  │   end
14  │   if x[node.feature]-node.threshold > 0 then
15  │   │   Queue.push(node.left, x[node.feature] - node.threshold)
16  │   │   Queue.push(node.right, 0)
17  │   else
18  │   │   Queue.push(node.left, 0)
19  │   │   Queue.push(node.right, node.threshold - x[node.feature])
20  │   end
21  end
22  return bestNode.instance
```

6.3 Classification Probability

For a classification task, K-nn can only return the label according to Eq. (6.1). In cases where the probability of classification is needed, there are ad hoc ways for this purpose. For example, we can let

$$P(c|\mathbf{x}) = \frac{k_c + s}{k + |C|s} \tag{6.8}$$

for all $c \in C$, where k_c is the number of instances in the k neighbors that are with label c, and s is a small positive constant that is used to avoid $P(c) = 0$ for those classes c that do not appear in the k neighbors.

Alternatively, let d_c be the average distance from $\mathbf{x}$ to the instances with label c in the k neighbors of $\mathbf{x}$, we can let

$$P(c|\mathbf{x}) = \frac{\exp(-d_c) + s}{\sum_{c \in C} \exp(-d_c) + |C|s} \tag{6.9}$$

Note that, we have $d_c = \infty$ (and thus $\exp(-d_c) = 0$) if there is no instance of label c in the k neighbors.

We remark that, the probability $P(c|\mathbf{x})$ in Eq. (6.8) is actually piece-wise linear over the changing of data instance $\mathbf{x}$, because it relies on the number of instances k_c. On the other hand, the one in Eq. (6.9) is smoother by considering the average distance d_c.

6.4   Robustness and Adversarial Attack

Given a dataset D of n input samples, and a number k, it is not hard to see that the input domain can be partitioned into $\binom{n}{k}$ disjoint partitions, each of which is determined by k samples such that all inputs in the partition take the k samples as their nearest neighbors. It is noted that, all samples in the same partition share the same predictive label. Moreover, each partition can be expressed as a finite set of $k(n - k)$ linear (in)equations, each of which expresses that it is closer to one of the k samples than any of the remaining $n - k$ samples. For any input $\mathbf{x}$, we write $R(\mathbf{x})$ for the partition it belongs to. Let $\mathcal{R}$ be the set of partitions, and we use r to range over $\mathcal{R}$.

Now, given an input sample $\mathbf{x}$ to be attacked, and any partition $r \neq R(\mathbf{x})$, we can define the following optimisation problem to find the input $\mathbf{x}'$ that is in partition r and closest to $\mathbf{x}$:

$$closest(\mathbf{x}, r) = \arg \min_{\mathbf{x}' \in r} ||\mathbf{x} - \mathbf{x}'|| \tag{6.10}$$

The problem can be solved by applying constraint solver over the above objective with the linear (in)equations of r as constraints. Based on this, the adversarial example can be obtained with the following optimisation problem:

$$r' = \arg \min_{r \in \mathcal{R}}[(\mathbf{x}' = closest(\mathbf{x}, r)) \wedge (y' \neq y)] \tag{6.11}$$

where y' and y are predictive labels of $\mathbf{x}'$ and $\mathbf{x}$, respectively. Finally, the optimal adversarial example is $closest(\mathbf{x}, r')$.

We remark that, the above method is suitable for any classifier that can partition the input domain into a finite number of partitions and each partition can be represented with a finite number of linear (in)equations, such as decision tree and random forest.

6.4.1   Is x′ an Adversarial Example?

Yes, the misclassification is ascertained by Eq. (6.11).

6.4.2   Is this Approach Complete?

Yes, the algorithm searches through all partitions (Eq. (6.11)) and all inputs in a partition (Eq. (6.10)).

6.4.3   Optimality

The resulting $\mathbf{x}'$ is optimal.

6.5   Poisoning Attack

We can use both the heuristic approaches and the alternating optimisation approach we will introduce in Sect. 10.8 for the poisoning attack. Heuristic approaches require the feature extraction function g, which is not available for K-nn. However, it can be replaced with the original sample, i.e., let $g(\mathbf{x}) = \mathbf{x}$, because K-nn is mostly working with low- or median-dimensional problems.

6.6   Model Stealing

K-nn algorithm does not have a model or trainable parameters. We can use the black-box algorithm that will be introduced in Sect. 10.9 to construct a functionally equivalent model.

6.7   Membership Inference

An example membership inference attack can be referred to the method in
Sect. 10.10. It is model agnostic and therefore can be applied to any machine
learning model.

6.8   Practice

For K-nn, we have the following code to **simpleML.py** which assumes $k = 3$.

```python
from sklearn import datasets
iris = datasets.load_iris()
X_train, X_test, y_train, y_test = train_test_split(iris.data,
    iris.target, test_size=0.20)

from sklearn.neighbors import KNeighborsClassifier
neigh = KNeighborsClassifier(n_neighbors=3)
neigh.fit(X_train, y_train)
print("Training accuracy is %s"% neigh.score(X_train,y_train))
print("Test accuracy is %s"% neigh.score(X_test,y_test))

print("Labels of all instances:\n%s"%y_test)
y_pred = neigh.predict(X_test)
print("Predictive outputs of all instances:\n%s"%y_pred)

from sklearn.metrics import classification_report,
    confusion_matrix
print("Confusion Matrix:\n%s"%confusion_matrix(y_test, y_pred))
print("Classification Report:\n%s"%classification_report(y_test,
    y_pred))

import numpy as np
from math import sqrt
from collections import Counter

# Implement kNN in details
def kNNClassify(K, X_train, y_train, X_predict):
    distances = [sqrt(np.sum((x - X_predict)**2)) for x in
    X_train]
    sort = np.argsort(distances)
    topK = [y_train[i] for i in sort[:K]]
    votes = Counter(topK)
    y_predict = votes.most_common(1)[0][0]
    return y_predict

def kNN_predict(K, X_train, y_train, X_predict, y_predict):
    acc = 0
    for i in range(len(X_predict)):
        if y_predict[i] == kNNClassify(K, X_train, y_train,
    X_predict[i]):
```

```
           acc += 1
     print(acc/len(X_predict))

print("Training accuracy is ", end='')
kNN_predict(3, X_train, y_train, X_train, y_train)
print("Test accuracy is ", end='')
kNN_predict(3, X_train, y_train, X_test, y_test)
```

```
import itertools
import copy

# Attack on KNN
def kNN_attack(K, X_train, y_train, X_predict, y_predict):
    m = np.diag([0.5,0.5,0.5,0.5])*4
    flag = True
    for i in range(1,5):
        for ii in list(itertools.combinations([0,1,2,3],i)):
            delta = np.zeros(4)
            for jj in ii:
                delta += m[jj]

            if y_predict != kNNClassify(K, X_train, y_train, copy
    .deepcopy(X_predict)+delta):
                X_predict += delta
                flag = False
                break

            if y_predict != kNNClassify(K, X_train, y_train, copy
    .deepcopy(X_predict)-delta):
                X_predict -= delta
                flag = False
                break
        if not flag:
            break

    print('attack data: ',X_predict)
    print('predict label: ',kNNClassify(K, X_train, y_train,
    X_predict))

X_test_ = X_test[0]
y_test_ = y_test[0]
print('original data: ', X_test_)
print('original label: ', y_test_)
kNN_attack(3, X_train, y_train, X_test_, y_test_)
```

Chapter 7
Linear Regression

Linear regression assumes a linear model as the underlying learning model. In the linear model, the output $\hat{y}$ can be expressed with a linear combination of its input features $X_1, \ldots, X_n$. While linear regression is less expressive (i.e., the hypothesis space is small), it has the significant benefit of being interpretable. Therefore, for either small applications (where linear models are sufficient for problems) or applications where interpretability goes over the precision, linear regression and its variants are popular machine learning algorithms. Moreover, linear regression has the other advantage of having an analytic solution, i.e., the optimal linear model can be learned.

In this section, we will first present linear regression and then discuss two variants: linear classification and logistic regression, in Sects. 7.2 and 7.3, respectively. We will then discuss robustness attack in Sect. 7.4, model stealing in Sect. 7.6, and other safety threats. Similar as the learning algorithm, some attacks of linear regression also have analytical solutions.

7.1 Linear Regression

Given a set of training data $D = \{(\mathbf{x}^{(i)}, y^{(i)}) | i \in \{1, \ldots, m\}\}$ sampled identically and independently from a distribution $\mathcal{D}$, linear regression assumes that the relationship between the label Y and the n-dimensional variable X is linear. More specifically, it is assume that the hypothesis space $\mathcal{H}$ is within linear models.

Therefore, the goal is to find a parameterised function

$$f_{\mathbf{w}}(x) = \mathbf{w}^T \mathbf{x} \tag{7.1}$$

© The Author(s), under exclusive license to Springer Nature Singapore Pte Ltd. 2023
X. Huang et al., *Machine Learning Safety*, Artificial Intelligence: Foundations, Theory, and Algorithms, https://doi.org/10.1007/978-981-19-6814-3_7

that minimises the following L_2 loss (or mean square error)

$$\hat{L}(f_{\mathbf{w}}) = \frac{1}{m}\sum_{i=1}^{m}(\mathbf{w}^T\mathbf{x}^{(i)} - y^{(i)})^2 \tag{7.2}$$

Note that, $\mathbf{w}^T\mathbf{x}^{(i)} - y^{(i)}$ represents the loss of the instance $\mathbf{x}^{(i)}$. If written in matrix form, it is

$$\hat{L}(f_{\mathbf{w}}) = \frac{1}{m}\sum_{i=1}^{m}(\mathbf{w}^T\mathbf{x}^{(i)} - y^{(i)})^2 = \frac{1}{m}||\mathbf{Xw} - \mathbf{y}||_2^2 \tag{7.3}$$

Example 7.1 Assume we have a set of four samples:

$$(\begin{bmatrix} 182 \\ 87 \\ 11.3 \end{bmatrix}, 325), (\begin{bmatrix} 189 \\ 92 \\ 12.3 \end{bmatrix}, 344), (\begin{bmatrix} 178 \\ 79 \\ 10.6 \end{bmatrix}, 350), (\begin{bmatrix} 183 \\ 90 \\ 12.7 \end{bmatrix}, 320) \tag{7.4}$$

Now, given a learned function $f_{\mathbf{w}}(\mathbf{x})$ such that $\mathbf{w} = (1, -1, 20)^T$, we have

$$\mathbf{w}^T\mathbf{x}^{(1)} = 321, \mathbf{w}^T\mathbf{x}^{(2)} = 343, \mathbf{w}^T\mathbf{x}^{(3)} = 311, \mathbf{w}^T\mathbf{x}^{(4)} = 347. \tag{7.5}$$

Therefore, we can compute the loss with Eq. (7.2).

7.1.1   Variant: Linear Regression with Bias

It is possible that we may consider the function f with bias term:

$$f_{\mathbf{w},\mathbf{b}}(x) = \mathbf{w}^T\mathbf{x} + \mathbf{b} \tag{7.6}$$

To handle this case, we can reduce it to the case without bias by letting

$$\mathbf{w}' = [\mathbf{w}; \mathbf{b}], \mathbf{x}' = [\mathbf{x}; 1] \tag{7.7}$$

Then, we have

$$f_{\mathbf{w},\mathbf{b}}(x) = \mathbf{w}^T\mathbf{x} + \mathbf{b} = (\mathbf{w}')^T(\mathbf{x}') \tag{7.8}$$

Intuitively, every instance is extended with one more feature whose value is always 1, and we assume that we already know the weight for this feature in $f_{\mathbf{w},\mathbf{b}}$, which is $\mathbf{b}$.

Example 7.2 Continue with Example 7.1, if we have $\mathbf{b} = (-330, -330, -330)^T$, we have

$$\mathbf{w}^T\mathbf{x}^{(1)} + b^{(1)} = -9, \mathbf{w}^T\mathbf{x}^{(2)} + b^{(2)} = 13, \mathbf{w}^T\mathbf{x}^{(3)} + b^{(3)}$$
$$= -19, \mathbf{w}^T\mathbf{x}^{(4)} + b^{(4)} = 17. \tag{7.9}$$

7.1.2 Variant: Linear Regression with Lasso Penalty

We may also be interested in adapting the loss, e.g., consider the following loss

$$\hat{L}(f_{\mathbf{w}}) = \frac{1}{m}\sum_{i=1}^{m}(\mathbf{w}^T\mathbf{x}^{(i)} - y^{(i)})^2 + \lambda||\mathbf{w}||_1 \tag{7.10}$$

where the lasso penalty term $||\mathbf{w}||_1$ is to encourage the sparsity of the weights.

7.2 Linear Classification

To consider classification task where $y^{(i)}$'s are labels instead of regression values, a natural attempt is to change the hypothesis class $\mathcal{H}$ from the set of linear models to a set of piece-wise linear models. For simplicity, assume that we are working with binary classification, i.e., $V(Y) = \{0, 1\}$. Then, the hypothesis class $\mathcal{H}$ is parameterised over $\mathbf{w}$ such that

$$f_{\mathbf{w}}(\mathbf{x}) = \begin{cases} 1 & \text{if } \mathbf{w}^T\mathbf{x} > 0 \\ 0 & \text{otherwise} \end{cases} \tag{7.11}$$

7.2.1 What is the Corresponding Optimisation Problem?

With the hypothesis class $\mathcal{H}$, the classification problem is to find a parameterised function

$$f'_{\mathbf{w}}(\mathbf{x}) = \mathbf{w}^T\mathbf{x} \tag{7.12}$$

that minimises the following loss

$$\hat{L}(f_{\mathbf{w}}) = \frac{1}{m} \sum_{i=1}^{m} I[step(\mathbf{w}^T \mathbf{x}^{(i)}), y^{(i)}] \qquad (7.13)$$

where *step* function is defined as: $step(m) = 1$ when $m > 0$ and $step(m) = 0$ otherwise, and $I(\hat{y}, y)$ is the 0-1 loss, i.e., $I(\hat{y}, y) = 0$ when $\hat{y} = y$ and $I(\hat{y}, y) = 1$ otherwise.

7.2.2 Difficulty of Finding Optimal Solution

Unfortunately, the finding of optimal solution to the optimisation problem in Eqs. (7.12) and (7.13) is NP-hard.

7.3 Logistic Regression

Given the above natural—but nevertheless naive—attempt does not necessarily lead to a good method for the classification problem, it is needed to consider other options, such as logistic regression.

Linear regression attacks the classification problem by attempting to make the regression values as probability values. That is, if the return of $f_{\mathbf{w}}(\mathbf{x})$ is a probability value of classifying $\mathbf{x}$ as 0, then we are easy to infer the classification by using the regression value of $\mathbf{x}$. This, however, is not straightforward as the linear regression may return values outside of [0,1].

7.3.1 Sigmoid Function

First of all, we need to make sure that the regression value within [0,1]. This can be done through applying the sigmoid function

$$\sigma(a) = \frac{1}{1 + \exp(-a)} \qquad (7.14)$$

which has the domain $(0, 1)$. Therefore, we can now update Eq. (7.13) into

$$\hat{L}(f_{\mathbf{w}}) = \frac{1}{m} \sum_{i=1}^{m} (\sigma(\mathbf{w}^T \mathbf{x}^{(i)}) - y^{(i)})^2 \qquad (7.15)$$

However, even so, it is still unclear whether $\sigma(\mathbf{w}^T\mathbf{x}^{(i)})$ represents the probability value.

7.3.2   *Force $\sigma(\mathbf{w}^T\mathbf{x}^{(i)})$ into Probability Value*

To achieve this, we make the following interpretation

$$P_{\mathbf{w}}(y = 1|\mathbf{x}) = \sigma(\mathbf{w}^T\mathbf{x}^{(i)})$$
$$P_{\mathbf{w}}(y = 0|\mathbf{x}) = 1 - P_{\mathbf{w}}(y = 1|\mathbf{x}) = 1 - \sigma(\mathbf{w}^T\mathbf{x}^{(i)}) \tag{7.16}$$

Then, we can update the loss (Eq. (7.13)) into

$$\hat{L}(f_{\mathbf{w}}) = -\frac{1}{m}\sum_{i=1}^{m}\log P_{\mathbf{w}}(y^{(i)}|\mathbf{x}^{(i)}) \tag{7.17}$$

By applying Eq. (7.16), we have

$$\begin{aligned}
\hat{L}(f_{\mathbf{w}}) &= -\frac{1}{m}\sum_{y^{(i)}=1}\log\sigma(\mathbf{w}^T\mathbf{x}^{(i)}) - \frac{1}{m}\sum_{y^{(i)}=0}\log[1-\sigma(\mathbf{w}^T\mathbf{x}^{(i)})] \\
&= -\frac{1}{m}\sum_{(\mathbf{x},y)\in D} y\log\sigma(\mathbf{w}^T\mathbf{x}) + (1-y)\log[1-\sigma(\mathbf{w}^T\mathbf{x})]
\end{aligned} \tag{7.18}$$

Example 7.3 Consider the four data samples in Example 7.1 (also provided in Table 7.1)

and the mean square error, if we have the following two functions:

- $f_{\mathbf{w}_1} = 2X_1 + 1X_2 + 20X_3 - 330$
- $f_{\mathbf{w}_2} = X_1 - 2X_2 + 23X_3 - 332$

Table 7.1 A small dataset

X_1	X_2	X_3	Y
182	87	11.3	325
189	92	12.3	344
178	79	10.6	350
183	90	12.7	320

(1) Because

$$
\begin{aligned}
f_{\mathbf{w}_1}(\mathbf{x}_1) - y_1 &= 2 * 182 + 1 * 87 + 20 * 11.3 - 330 - 325 = 22 \\
f_{\mathbf{w}_1}(\mathbf{x}_2) - y_2 &= 2 * 189 + 1 * 92 + 20 * 12.3 - 330 - 344 = 42 \\
f_{\mathbf{w}_1}(\mathbf{x}_3) - y_3 &= 2 * 178 + 1 * 79 + 20 * 10.6 - 330 - 350 = -33 \\
f_{\mathbf{w}_1}(\mathbf{x}_4) - y_4 &= 2 * 183 + 1 * 90 + 20 * 12.7 - 330 - 320 = 60 \\
f_{\mathbf{w}_2}(\mathbf{x}_1) - y_1 &= 182 - 2 * 87 + 23 * 11.3 - 332 - 325 \quad = -389.1 \\
f_{\mathbf{w}_2}(\mathbf{x}_2) - y_2 &= 189 - 2 * 92 + 23 * 12.3 - 332 - 344 \quad = -388.1 \\
f_{\mathbf{w}_2}(\mathbf{x}_3) - y_3 &= 178 - 2 * 79 + 23 * 10.6 - 332 - 350 \quad = -418.2 \\
f_{\mathbf{w}_2}(\mathbf{x}_4) - y_4 &= 183 - 2 * 90 + 23 * 12.7 - 332 - 320 \quad = -356.9
\end{aligned}
\tag{7.19}
$$

the model $f_{\mathbf{w}_1}$ is better than $f_{\mathbf{w}_2}$ for linear regression, according to the loss function (Eq. (7.3));

(2) For an instance $\mathbf{y}^T = (0, 1, 1, 0)$, because

$$
\begin{aligned}
step(f_{\mathbf{w}_1}(\mathbf{x}_1)) &= 1 \\
step(f_{\mathbf{w}_1}(\mathbf{x}_2)) &= 1 \\
step(f_{\mathbf{w}_1}(\mathbf{x}_3)) &= 1 \\
step(f_{\mathbf{w}_1}(\mathbf{x}_4)) &= 1 \\
step(f_{\mathbf{w}_2}(\mathbf{x}_1)) &= 0 \\
step(f_{\mathbf{w}_2}(\mathbf{x}_2)) &= 0 \\
step(f_{\mathbf{w}_2}(\mathbf{x}_3)) &= 0 \\
step(f_{\mathbf{w}_2}(\mathbf{x}_4)) &= 0
\end{aligned}
\tag{7.20}
$$

we have that both models are the same for linear classification by considering 0-1 loss, according to the loss function (Eq. (7.13));

(3) For an instance $\mathbf{y}^T = (0, 1, 1, 0)$, because

$$
\begin{aligned}
y_1 * \log(\sigma(f_{\mathbf{w}_1}(\mathbf{x}_1))) + (1 - y_1) \log((1 - \sigma(f_{\mathbf{w}_1}(\mathbf{x}_1)))) &= -M \\
y_2 * \log(\sigma(f_{\mathbf{w}_1}(\mathbf{x}_2))) + (1 - y_2) \log((1 - \sigma(f_{\mathbf{w}_1}(\mathbf{x}_2)))) &= 0 \\
y_3 * \log(\sigma(f_{\mathbf{w}_1}(\mathbf{x}_3))) + (1 - y_3) \log((1 - \sigma(f_{\mathbf{w}_1}(\mathbf{x}_3)))) &= 0 \\
y_4 * \log(\sigma(f_{\mathbf{w}_1}(\mathbf{x}_4))) + (1 - y_4) \log((1 - \sigma(f_{\mathbf{w}_1}(\mathbf{x}_4)))) &= -M \\
y_1 * \log(\sigma(f_{\mathbf{w}_2}(\mathbf{x}_1))) + (1 - y_1) \log((1 - \sigma(f_{\mathbf{w}_2}(\mathbf{x}_1)))) &= 0 \\
y_2 * \log(\sigma(f_{\mathbf{w}_2}(\mathbf{x}_2))) + (1 - y_2) \log((1 - \sigma(f_{\mathbf{w}_2}(\mathbf{x}_2)))) &= -44.1 \\
y_3 * \log(\sigma(f_{\mathbf{w}_2}(\mathbf{x}_3))) + (1 - y_3) \log((1 - \sigma(f_{\mathbf{w}_2}(\mathbf{x}_3)))) &= -68.2 \\
y_4 * \log(\sigma(f_{\mathbf{w}_2}(\mathbf{x}_4))) + (1 - y_4) \log((1 - \sigma(f_{\mathbf{w}_2}(\mathbf{x}_4)))) &= -1.1
\end{aligned}
\tag{7.21}
$$

where M represents a large number, so we have

$$
\begin{aligned}
\hat{L}(f_{\mathbf{w}_1}) &= -\tfrac{1}{4}(-M + 0 + 0 - M) = M/2 \\
\hat{L}(f_{\mathbf{w}_2}) &= -\tfrac{1}{4}(0 - 44.1 - 68.2 - 1.1) = 28.35
\end{aligned}
\tag{7.22}
$$

according to Eq. (7.18). Therefore, $f_{\mathbf{w}_2}$ is better for logistic regression.

(4) According to the logistic regression of the first model, to know the prediction result of the first model on a new input $(181, 92, 12.4)$, according to Eq. (7.16), we have

$$P_{\mathbf{w}_1}(y = 1|\mathbf{x}) = \sigma(2 * 181 + 1 * 92 + 20 * 12.4 - 330) = 1$$
$$P_{\mathbf{w}_1}(y = 0|\mathbf{x}) = 1 - \sigma(2 * 181 + 1 * 92 + 20 * 12.4 - 330) = 0 \qquad (7.23)$$

Therefore, it is predicted to 1.

7.4 Robustness and Adversarial Attack

Considering a Binary classification problem, i.e., $y \in \{+1, -1\}$ for any data instance $\mathbf{x}$, the loss for a single instance $(\mathbf{x}, y)$ as in Eq. (7.18) is

$$\hat{L}(f_{\mathbf{w}}, (\mathbf{x}, y)) = -y \log \sigma(\mathbf{w}^T \mathbf{x}) - (1 - y) \log[1 - \sigma(\mathbf{w}^T \mathbf{x})] \qquad (7.24)$$

The computation of adversarial example is to solve the following optimisation problem:

$$\text{maximise}_{||\delta|| \le \epsilon} \hat{L}(f_{\mathbf{w}}, (\mathbf{x} + \delta, y)) \qquad (7.25)$$

which finds the greatest loss within the constraint over the perturbation δ. We note that, by Eq. (7.24), this is equivalent to

$$\text{minimise}_{||\delta|| \le \epsilon} y\mathbf{w}^T \delta \qquad (7.26)$$

Now, considering the L_∞ norm, i.e., $||\delta||_\infty \le \epsilon$, the optimal solution to Eq. (7.26) is

$$\delta^* = -y\epsilon sign(\mathbf{w}) \qquad (7.27)$$

where $sign$ is the sign function extracting the sign of real numbers, such that

$$sign(w) = \begin{cases} -1 & w < 0 \\ 0 & w = 0 \\ 1 & w > 0 \end{cases} \qquad (7.28)$$

and $sign(\mathbf{w})$ is the application of $sign$ on all entries of $\mathbf{w}$.

7.5   Poisoning Attack

We can use both the heuristic approaches and the alternating optimisation approach we will introduce in Sect. 10.8 for the poisoning attack. Heuristic approaches require the feature extraction function g, which can be replaced with the original sample, i.e., let $g(\mathbf{x}) = \mathbf{x}$.

7.6   Model Stealing

Recall that model stealing attack is to reconstruct another model f' given an existing, trained model f. In this section, we consider a simplified scenario where f and f' share the same structure with the only difference being on the trainable parameters.

For the case of linear regression with regularisation (such as the lasso penalty as in Eq. (7.10)), we write the loss function as

$$\hat{L}(f_{\mathbf{w}}) = L(\mathbf{X}, \mathbf{y}; \mathbf{w}) + \lambda R(\mathbf{w}) \tag{7.29}$$

where $L(\mathbf{X}, \mathbf{y}; \mathbf{w})$ measures the loss of the dataset when the linear model takes the parameters $\mathbf{w}$, and $R(\mathbf{w})$ is the regularisation term.

Assume that, for model stealing attack, we want to learn both $\mathbf{w}'$ and λ'. Moreover, we sample another dataset $\mathbf{X}'$ with its prediction $\mathbf{y}'$. Similarly as the learning algorithm, we let the gradient of the objective function $\hat{L}(f_{\mathbf{w}})$ be $\mathbf{0}$, i.e.,

$$\frac{\partial \hat{L}(f_{\mathbf{w}})}{\partial \mathbf{w}} = \mathbf{b} + \lambda \mathbf{a} = \mathbf{0} \tag{7.30}$$

where

$$\begin{aligned} \mathbf{b} &= [\frac{\partial L(\mathbf{X}, \mathbf{y}; \mathbf{w})}{\partial w_1}, \dots, \frac{\partial L(\mathbf{X}, \mathbf{y}; \mathbf{w})}{\partial w_{|\mathbf{w}|}}]^T \\ \mathbf{a} &= [\frac{R(\mathbf{w})}{\partial w_1}, \dots, \frac{R(\mathbf{w})}{\partial w_{|\mathbf{w}|}}]^T \end{aligned} \tag{7.31}$$

Note that, Eq. (7.30) is a system of linear equations, where the number of equations can be more than the number of variables. In such case, we can estimate

$$\lambda = -(\mathbf{a}^T \mathbf{a})^{-1} \mathbf{a}^T \mathbf{b} \tag{7.32}$$

7.7 Membership Inference

A few examples of membership inference attacks can be referred to Sect. 10.10. Some of them are model agnostic and therefore can be applied to any machine learning model.

7.8 Practice

For logistic regression, we add the following code to the **simpleML.py** file:

```
from sklearn import datasets
iris = datasets.load_iris()
X_train, X_test, y_train, y_test = train_test_split(iris.data
    [:100], iris.target[:100], test_size=0.20)
```

```
from sklearn.linear_model import LogisticRegression
reg = LogisticRegression(solver='lbfgs', max_iter=10000)
reg.fit(X_train, y_train)
print("Training accuracy is %s"% reg.score(X_train,y_train))
print("Test accuracy is %s"% reg.score(X_test,y_test))
```

```
print("Labels of all instances:\n%s"%y_test)
y_pred = reg.predict(X_test)
print("Predictive outputs of all instances:\n%s"%y_pred)

from sklearn.metrics import classification_report,
    confusion_matrix
print("Confusion Matrix:\n%s"%confusion_matrix(y_test, y_pred))
print("Classification Report:\n%s"%classification_report(y_test,
    y_pred))
```

```
import numpy as np
from sklearn.metrics import accuracy_score
def sigmoid(x):
    z = 1 / (1 + np.exp(-x))
    return z

def add_b(dataMatrix):
    dataMatrix = np.column_stack((np.mat(dataMatrix),np.ones(np.
    shape(dataMatrix)[0])))
    return dataMatrix

def LogisticRegression_(x_train,y_train,x_test,y_test,alpha =
    0.001 ,maxCycles = 500):
    x_train = add_b(x_train)
    x_test = add_b(x_test)
    y_train = np.mat(y_train).transpose()
    y_test = np.mat(y_test).transpose()
    m,n = np.shape(x_train)
```

```
17    weights = np.ones((n,1))
18    for i in range(0,maxCycles):
19        h = sigmoid(x_train*weights)
20        error = y_train - h
21        weights = weights + alpha * x_train.transpose() * error
22
23    y_pre = sigmoid(np.dot(x_train, weights))
24    for i in range(len(y_pre)):
25        if y_pre[i] > 0.5:
26            y_pre[i] = 1
27        else:
28            y_pre[i] = 0
29    print("Train accuracy is %s"% (accuracy_score(y_train, y_pre)
      ))
30
31    y_pre = sigmoid(np.dot(x_test, weights))
32    for i in range(len(y_pre)):
33        if y_pre[i] > 0.5:
34            y_pre[i] = 1
35        else:
36            y_pre[i] = 0
37    print("Test accuracy is %s"% (accuracy_score(y_test, y_pre)))
38
39 weights = LogisticRegression_(X_train, y_train,X_test,y_test)
```

```
1  import itertools
2  import copy
3
4  # Attack on LogisticRegression
5  def LogisticRegression_attack(weights, X_predict, y_predict):
6      X_predict = add_b(X_predict)
7      m = np.diag([0.5,0.5,0.5,0.5])*4
8      flag = True
9      for i in range(1,5):
10         for ii in list(itertools.combinations([0,1,2,3],i)):
11             delta = np.zeros(4)
12             for jj in ii:
13                 delta += m[jj]
14             delta = np.append(delta, 0.)
15
16             y_pre = sigmoid(np.dot(copy.deepcopy(X_predict)+delta
      , weights))
17             if y_pre > 0.5:
18                 y_pre = 1
19             else:
20                 y_pre = 0
21             if y_predict != y_pre:
22                 X_predict += delta
23                 flag = False
24                 break
25
26             y_pre = sigmoid(np.dot(copy.deepcopy(X_predict)-delta
      , weights))
27             if y_pre > 0.5:
```

```
28            y_pre = 1
29          else:
30              y_pre = 0
31          if y_predict != y_pre:
32              X_predict -= delta
33              flag = False
34              break
35      if not flag:
36          break

38  y_pre = sigmoid(np.dot(X_predict, weights))
39  if y_pre > 0.5:
40      y_pre = 1
41  else:
42      y_pre = 0
43  print('attack data: ', X_predict[0,:-1])
44  print('predict label: ', y_pre)

46 X_test_ = X_test[0:1]
47 y_test_ = y_test[0]
48 print('original data: ', X_test_)
49 print('original label: ', y_test_)
50 LogisticRegression_attack(weights, X_test_, y_test_)
```

Chapter 8
Naive Bayes

Naive Bayes is a probabilistic algorithm that can be used for classification problems. Albeit simple and intuitive, naive Bayes performs very well in many practical applications such as the spam filter email application. In the following, we will first introduce the learning algorithm, and then discuss how the naive Bayes may be attacked with respect to the safety properties we discussed in Chap. 3.

8.1 Learning Algorithm

Let Y be a random variable representing the label and $X_1, \ldots, X_n$ be random variables representing the n input features, respectively. The classification problem can be expressed as

$$\arg\max_{y \in V(Y)} P(Y = y | X_1 = x_1, \ldots, X_n = x_n) \qquad (8.1)$$

which is to find the label y with the maximum probability given the instance $\mathbf{x} = (x_1, \ldots, x_n)$. This can be done by first estimating the following conditional probability table from the training dataset D

$$P(Y | X_1, \ldots, X_n) \qquad (8.2)$$

and then apply Eq. (8.1) on $\mathbf{x}$.

Bayes Theorem

Bayes theorem suggests that we have

© The Author(s), under exclusive license to Springer Nature Singapore Pte Ltd. 2023
X. Huang et al., *Machine Learning Safety*, Artificial Intelligence: Foundations, Theory, and Algorithms, https://doi.org/10.1007/978-981-19-6814-3_8

$$P(Y|X) = \frac{P(X|Y)P(Y)}{P(X)} \tag{8.3}$$

where $P(Y)$ is the prior, $P(X)$ is the evidence, $P(X|Y)$ is the likelihood function, and $P(Y|X)$ is the posterior. By Bayes theorem, we have that

$$P(Y|X_1, \ldots, X_n) = \frac{P(X_1, \ldots, X_n|Y)P(Y)}{P(X_1, \ldots, X_n)} \tag{8.4}$$

That is, the computation of the conditional probability table $P(Y|X_1, \ldots, X_n)$ can be reduced to the computation of three tables $P(X_1, \ldots, X_n|Y)$, $P(X_1, \ldots, X_n)$, and $P(Y)$. Furthermore, noting that $P(X_1, \ldots, X_n)$—representing the data distribution—is fixed for any $y \in V(Y)$, we have that

$$\arg\max_{y \in V(Y)} P(Y = y|X_1 = x_1, \ldots, X_n = x_n)$$
$$\propto \arg\max_{y \in V(Y)} P(X_1 = x_1, \ldots, X_n = x_n|Y = y)P(Y = y) \tag{8.5}$$

Therefore, for the classification problem, it is sufficient to compute two probability tables: $P(X_1, \ldots, X_n|Y)$ and $P(Y)$.

Estimation of $P(Y)$

Estimation of $P(Y)$ can be done by letting

$$P(Y = y) = \frac{\text{Number of instances whose label is } y}{\text{Number of all instances}} \tag{8.6}$$

for all $y \in V(Y)$. Can we use similar expression to estimate $P(X_1, \ldots, X_n|Y)$? Yes, we can, but it is not scalable.

Difficulty of Estimating $P(X_1, \ldots, X_n|Y)$ Directly

Without loss of generality, we assume that all random variables are Boolean. Therefore, to estimation $P(Y)$, we need only compute once the Expression (8.6). If we want to estimate $P(X_1, \ldots, X_n|Y)$ with a similar expression as Expression (8.6), we will need $(2^n - 1) \times 2$ computations—an exponential computation.

Assumption

Naive Bayes assumes that the input features $X_1, \ldots, X_n$ are conditionally independent given the label Y, i.e.,

$$P(X_1, \ldots, X_n | Y) = \prod_{i=1}^{n} P(X_i | Y) \tag{8.7}$$

With this assumption, $P(X_1, \ldots, X_n | Y)$ can be estimated by computing n tables $P(X_i | Y)$, each of which requires 2 computations. That is, instead of conducting $(2^n - 1) \times 2$ computations, we now need $2n$ computations—a linear time computation.

With Assumption (8.7), we have the Naive Bayes expression:

$$\begin{aligned} &\arg\max_{y \in V(Y)} P(Y = y | X_1 = x_1, \ldots, X_n = x_n) \\ &\propto \arg\max_{y \in V(Y)} P(Y = y) \prod_{i=1}^{n} P(X_i = x_i | Y = y) \end{aligned} \tag{8.8}$$

Algorithm

The Naive Bayes classification algorithm proceeds in the following three steps:

1. For each value $y_k \in V(Y)$, we estimate

$$\pi_k = P(Y = y_k) \tag{8.9}$$

2. For each value x_{ij} of each feature X_i and each $y_k \in V(Y)$, we estimate

$$\theta_{ijk} = P(X_i = x_{ij} | Y = y_k) \tag{8.10}$$

3. Given a new input $\mathbf{x}_{new} = (x_1^{new}, \ldots, x_n^{new})$, we classify it by letting

$$\hat{y_{new}} \leftarrow \arg\max_{y_k} P(Y = y_k) \prod_i P(X_i = x_i^{new} | Y = y_k) \tag{8.11}$$

For Continuous Features

The above works with categorical features. For continuous features, we can discretise the values of the feature into a set of categorical values, so that the above method works. We may also consider making assumptions about the distribution of the continuous features.

For example, let $P(X_i|Y)$ be a Gaussian probability density function, i.e.,

$$P(X_i|Y) = \frac{1}{\sigma\sqrt{2\pi}} \exp(\frac{-(X_i - \mu)^2}{2\sigma^2}) \tag{8.12}$$

such that the Gaussian distribution has the expected value μ and variance σ^2. We can estimate μ with

$$\mu = \frac{\sum_{(\mathbf{x},y)\in D} X_i(\mathbf{x})}{|D|} \tag{8.13}$$

where $X_i(\mathbf{x})$ returns the value of feature X_i in instance $\mathbf{x}$, and σ^2

$$\sigma^2 = \frac{1}{|D| - 1} \sum_{(\mathbf{x},y)\in D} (X_i(\mathbf{x}) - \mu)^2 \tag{8.14}$$

Then we can have compute $P(X_i = x_i^{new}|Y = y_k)$ by applying Eq. (8.12). Afterwards, we can get the classification by Eq. (8.11).

8.2   Robustness and Adversarial Attack

The following is a heuristic algorithm. based on the definition of probability as in Eq. (8.11). Assume that we have a data instance with label $(\mathbf{x}, y)$ and want to find another one $(\mathbf{x}', y')$ that is close to $\mathbf{x}$. The general idea is to move the instance $\mathbf{x}$ gradually along the direction where the probability of being classified as y, i.e., $\prod_i P(X_i = x_i^{new}|Y = y_k)$, decreases the most, until the class change.

First of all, we define

$$\mathbf{x}_i^s = \begin{cases} \mathbf{x} + \epsilon_i & \text{if } s = + \\ \mathbf{x} - \epsilon_i & \text{if } s = - \end{cases} \tag{8.15}$$

where ϵ_i is a 0-vector except for the entry for feature X_i, which has value $\epsilon > 0$. Then, the gradient along the direction defined by X_i and $s \in \{+, -\}$ is as follows:

$$\nabla_i^s f(\mathbf{x}) = \frac{f(\mathbf{x}_i^s) - f(\mathbf{x})}{||\mathbf{x} - \mathbf{x}_i^s||} \tag{8.16}$$

Then, for the next step, we move $\mathbf{x}$ along the following direction to $\mathbf{x}'$:

$$\arg\min_{i,s} \nabla_i^s f(\mathbf{x}) \tag{8.17}$$

and check if there is a misclassification on $\mathbf{x}'$. We repeat the above move until there is a misclassification.

We remark that, the above method is model agnostic, i.e., it can work with any machine learning model as long as the attacker is able to query the model to obtain the predictive probability for an input.

8.2.1 Is x' an Adversarial Example?

Yes, because the final $\mathbf{x}'$ is obtained by following a sequence of changes until witnessing a misclassification.

8.2.2 Is this Approach Complete?

Unfortunately, the above algorithm is incomplete.

8.2.3 Sub-Optimality

The resulting $\mathbf{x}'$ does not necessarily be the optimal solution.

8.3 Poisoning Attack

As observed from Eq. (8.11), whether an input $\mathbf{x}_{new}$ is classified as a label $\hat{y}_{new}$ is dependent on the appearance probability of the features x_i^{new} when the class is $\hat{y}_{new}$. Therefore, when intending to force the labelling of $\mathbf{x}_{adv}$ as y_{new}, a simple data poisoning attack is to add into the training dataset many poisoned samples $(\mathbf{x}, y)$ such that features of $\mathbf{x}_{adv}$ are likely to appear in the training dataset and y_{adv} is the target label. By doing so, when we train a Naive Bayes classifier with this poisoned dataset, the features in the poisoned samples have a higher probability when the class is y, which will lead to a higher probability of classifying future inputs with these features as y.

We can also use both the heuristic approaches and the alternating optimisation approach we will introduce in Sect. 10.8 for the poisoning attack. Heuristic approaches require the feature extraction function g, which is not available for Naive Bayes classifier. However, it can be replaced with the original sample, i.e., let $g(\mathbf{x}) = \mathbf{x}$.

8.4 Model Stealing

Naive Bayes algorithm does not have model or trainable parameters. We can use the black-box algorithm that will be introduced in Sect. 10.9 to construct a functionally equivalent model.

8.5 Membership Inference

A few examples of membership inference attacks can be referred to Sect. 10.10. Some of them are model agnostic and therefore can be applied to any machine learning model.

8.6 Model Inversion

We follow the discussion in Sect. 5.8 to use the formalisation of the problem in Eq. (5.19). For Naive Bayes, it can be transformed as follows.

$$
\begin{aligned}
& \arg\max_{v \in V(X_1)} P(X_1 = v \mid X_{-1} = \mathbf{x}_{-1}, y) \\
& = \arg\max_{v \in V(X_1)} \frac{P(y)P(v, \mathbf{x}_{-1}|y)}{P(y)P(\mathbf{x}_{-1}|y)} \\
& = \arg\max_{v \in V(X_1)} \frac{P(v|y)P(\mathbf{x}_{-1}|y)}{P(\mathbf{x}_{-1}|y)} = \arg\max_{v \in V(X_1)} P(v|y)
\end{aligned}
\tag{8.18}
$$

Then, this can be estimated with a set of randomly sampled data instances.

8.7 Practice

We can use Gaussian naive Bayes to **simpleML.py** as follows:

```
from sklearn import datasets
iris = datasets.load_iris()
X_train, X_test, y_train, y_test = train_test_split(iris.data,
    iris.target, test_size=0.20)
```

```
from sklearn.naive_bayes import GaussianNB
gnb = GaussianNB()
gnb.fit(X_train, y_train)
print("Training accuracy is %s"% gnb.score(X_train,y_train))
print("Test accuracy is %s"% gnb.score(X_test,y_test))
```

```
print("Labels of all instances:\n%s"%y_test)
y_pred = gnb.predict(X_test)
print("Predictive outputs of all instances:\n%s"%y_pred)

from sklearn.metrics import classification_report,
    confusion_matrix
print("Confusion Matrix:\n%s"%confusion_matrix(y_test, y_pred))
print("Classification Report:\n%s"%classification_report(y_test,
    y_pred))
```

```
from collections import Counter

class GaussianNB_:
    def __init__(self):
        self.prior = None
        self.avgs = None
        self.vars = None
        self.n_class = None

    def _get_prior(self, y):
        cnt = Counter(y)
        prior = np.array([cnt[i] / len(y) for i in range(len(cnt)
    )])
        return prior

    def _get_avgs(self, X, y):
        return np.array([X[y == i].mean(axis=0) for i in range(
    self.n_class)])

    def _get_vars(self, X, y):
        return np.array([X[y == i].var(axis=0) for i in range(
    self.n_class)])

    def _get_likelihood(self, row):
        return (1 / np.sqrt(2 * np.pi * self.vars) * np.exp(-(row
    - self.avgs)**2 / (2 * self.vars))).prod(axis=1)

    def fit(self, X, y):
        self.prior = self._get_prior(y)
        self.n_class = len(self.prior)
        self.avgs = self._get_avgs(X, y)
        self.vars = self._get_vars(X, y)

    def predict_prob(self, X):
        likelihood = np.apply_along_axis(self._get_likelihood,
    axis=1, arr=X)
        probs = self.prior * likelihood
        probs_sum = probs.sum(axis=1)
        return probs / probs_sum[:, None]

    def predict(self, X):
        return self.predict_prob(X).argmax(axis=1)

def get_acc(y, y_hat):
```

```
40      a = 0
41      for i in range(len(y)):
42          if y[i]==y_hat[i]:
43              a += 1
44      return a/len(y)
45
46  clf = GaussianNB_()
47  clf.fit(X_train, y_train)
48
49  y_hat = clf.predict(X_train)
50  acc = get_acc(y_train, y_hat)
51  print("Train accuracy is %s"% acc)
52
53  y_hat = clf.predict(X_test)
54  acc = get_acc(y_test, y_hat)
55  print("Test accuracy is %s"% acc)
```

```
1   import itertools
2
3   # Attack on GaussianNB
4   def GaussianNB_attack(clf, X_predict, y_predict):
5       m = np.diag([0.5,0.5,0.5,0.5])*4
6       flag = True
7       for i in range(1,5):
8           for ii in list(itertools.combinations([0,1,2,3],i)):
9               delta = np.zeros(4)
10              for jj in ii:
11                  delta += m[jj]
12
13              y_pre = clf.predict(copy.deepcopy(X_predict)+delta)
14              if y_predict != y_pre:
15                  X_predict += delta
16                  flag = False
17                  break
18
19              y_pre = clf.predict(copy.deepcopy(X_predict)-delta)
20              if y_predict != y_pre:
21                  X_predict -= delta
22                  flag = False
23                  break
24          if not flag:
25              break
26
27      print('attack data: ', X_predict)
28      print('predict label: ', clf.predict(copy.deepcopy(X_predict)
        ))
29
30  X_test_ = X_test[0:1]
31  y_test_ = y_test[0]
32  print('original data: ', X_test_)
33  print('original label: ', y_test_)
34  GaussianNB_attack(clf, X_test_, y_test_)
```

Chapter 9
Loss Function and Gradient Descent

Before proceeding to deep learning, we use this section to discuss two key concepts: loss function and gradient descent. Technically, machine learning is to optimise a certain loss function over a set of training instances. A carefully designed loss function can significantly improve the performance of the trained model. A recent trend in machine learning research—as we will explain later in Part III—also designs loss functions to integrate safety properties. Once the loss function is designed, it is of our interest to find an optimal solution (i.e., a trained model) with respect to the loss function. However, the optimal solution might not be easily achievable, in particular for high-dimensional problems and/or for machine learning models with many parameters. In this case, gradient descent (and its variant such as stochastic gradient descent) is an effective method to find sub-optimal, yet often satisfactory, solutions. This chapter introduces the basic ideas of both loss function and gradient descent, and explains how they are applied to the case of linear regression.

9.1 Loss Functions

This section introduces variants of loss functions, which are learning objectives. Assume that, we have a model $f_{\mathbf{W}}$, being it a model whose parameters are just initialised or a model who appears during the training process. The dataset is $D = \{(\mathbf{x}_i, y_i) \mid i \in \{1..n\}\}$ is a labelled dataset. We are considering the classification task.

In previous sections, we have introduced mean squared error (MSE), repeated as below, for linear regression. MSE is one of the most widely used loss functions.

$$\hat{L}(f_{\mathbf{W}}) = \frac{1}{m} \sum_{i=1}^{m} (f_{\mathbf{W}}(\mathbf{x}^{(i)}) - y^{(i)})^2 \tag{9.1}$$

© The Author(s), under exclusive license to Springer Nature Singapore Pte Ltd. 2023
X. Huang et al., *Machine Learning Safety*, Artificial Intelligence: Foundations,
Theory, and Algorithms, https://doi.org/10.1007/978-981-19-6814-3_9

Intuitively, MSE measures the average areas of the square created by the predicted and ground-truth points.

9.1.1   Mean Absolute Error

$$\hat{L}(f_{\mathbf{w}}) = \frac{1}{m} \sum_{i=1}^{m} |f_{\mathbf{w}}(\mathbf{x}^{(i)}) - y^{(i)}| \tag{9.2}$$

Unlike MSE which concerns the areas of square, MAE concerns the geometrical distance between the predicted and ground-truth points. Comparing to MSE whose derivative can be easily computed, it is harder to compute derivative for MAE.

9.1.2   Root Mean Squared Error (RMSE)

RMSE is very similar to MSE, except for the square root operation.

$$\hat{L}(f_{\mathbf{w}}) = \sqrt{\frac{1}{m} \sum_{i=1}^{m} (f_{\mathbf{w}}(\mathbf{x}^{(i)}) - y^{(i)})^2} \tag{9.3}$$

9.1.3   Binary Cross Entropy Cost Function

When considering binary classification, i.e., $C = \{0, 1\}$, we may utilise information theoretical concepts, cross entropy, which measures the difference between two distributions for the predictions and the ground truths.

$$\hat{L}(f_{\mathbf{w}}) = \sum_{i=1}^{m} -y^{(i)} \log f_{\mathbf{w}}(\mathbf{x}^{(i)}) - (1 - y^{(i)}) \log(1 - f_{\mathbf{w}}(\mathbf{x}^{(i)})) \tag{9.4}$$

where Cross entropy loss works better after the softmax layer, because the output of the softmax layer represents a distribution.

9.1.4   Categorical Cross Entropy Cost Function

Extending the above to multiple classes, we may have

$$\hat{L}(f_{\mathbf{w}}) = -\sum_{i=1}^{m}\sum_{c\in C}\mathbf{y}_{c}^{(i)}\log[f_{\mathbf{w}}(\mathbf{x}^{(i)})]_{c} \tag{9.5}$$

where $\mathbf{y}^{(i)}$ is the one-hot representation of the ground truth $y^{(i)}$ and $\mathbf{y}_{c}^{(i)}$ denotes the component of $\mathbf{y}^{(i)}$ that is for the class c. Also, unlike the previous notations, $f_{\mathbf{w}}(\mathbf{x}^{(i)})$ is a probability distribution of the prediction over $\mathbf{x}^{(i)}$, and $[f_{\mathbf{w}}(\mathbf{x}^{(i)})]_{c}$ denotes the component of $f_{\mathbf{w}}(\mathbf{x}^{(i)})$ that is for the class c.

9.2   Gradient Descent

9.2.1   Derivative of a Function

Given a function $\hat{L}(x)$, its derivative is the slope of $\hat{L}(x)$ at point x, written as $\hat{L}'(x)$. It specifies how to scale a small change in input to obtain a corresponding change in the output. Specifically,

$$\hat{L}(x+\epsilon) \approx \hat{L}(x) + \epsilon\hat{L}'(x) \tag{9.6}$$

Moreover, define the sign function as the following:

$$sign(x) = \begin{cases} -1 & \text{if } x < 0 \\ 0 & \text{if } x = 0 \\ 1 & \text{otherwise} \end{cases} \tag{9.7}$$

Then, we have that

$$\hat{L}(x - \epsilon sign(\hat{L}'(x))) < \hat{L}(x) \tag{9.8}$$

Therefore, we can reduce $\hat{L}(x)$ by moving x in small steps with an opposite sign of derivative.

9.2.2   Gradient

Gradient generalizes notion of derivative where derivative is with respect to a vector.

$$\nabla_{\mathbf{x}}(\hat{L}(\mathbf{x})) = (\frac{\partial \hat{L}(\mathbf{x})}{\partial x_1}, \ldots, \frac{\partial \hat{L}(\mathbf{x})}{\partial x_n}) \tag{9.9}$$

where the partial derivative $\dfrac{\partial \hat{L}(\mathbf{x})}{\partial x_i}$ measures how the function $\hat{L}$ changes when only variable x_i increases at point $\mathbf{x}$.

9.2.3   Critical Points

Critical points are where every element of the gradient is equal to zero, i.e.,

$$\nabla_{\mathbf{x}}(\hat{L}(\mathbf{x})) = 0 \equiv \begin{cases} \dfrac{\partial \hat{L}(\mathbf{x})}{\partial x_1} = 0 \\ \ldots \\ \dfrac{\partial \hat{L}(\mathbf{x})}{\partial x_n} = 0 \end{cases} \tag{9.10}$$

9.2.4   Gradient Descent on Linear Regression

Given Eq. (7.3), we have that

$$\begin{aligned} &\nabla_{\mathbf{w}}\hat{L}(f_{\mathbf{w}}) \\ &= \nabla_{\mathbf{w}}\frac{1}{m}||\mathbf{Xw} - \mathbf{y}||_2^2 \\ &= \nabla_{\mathbf{w}}[(\mathbf{Xw} - \mathbf{y})^T(\mathbf{Xw} - \mathbf{y})] \\ &= \nabla_{\mathbf{w}}[\mathbf{w}^T\mathbf{X}^T\mathbf{Xw} - 2\mathbf{w}^T\mathbf{X}^T\mathbf{y} + \mathbf{y}^T\mathbf{y}] \\ &= 2\mathbf{X}^T\mathbf{Xw} - 2\mathbf{X}^T\mathbf{y} \end{aligned} \tag{9.11}$$

Therefore, we can follow the following gradient descent algorithm to solve linear regression:

1. Set step size ϵ and tolerance δ to small positive numbers
2. While $||\mathbf{X}^T\mathbf{Xw} - \mathbf{X}^T\mathbf{y}||_2 > \delta$ do

$$\mathbf{x} \leftarrow \mathbf{x} - \epsilon(\mathbf{X}^T\mathbf{Xw} - \mathbf{X}^T\mathbf{y}) \tag{9.12}$$

3. Return $\mathbf{x}$ as a solution

9.2.5   *Analytical Solution on Linear Regression*

We may be able to avoid iterative algorithm and jump to the critical point by solving the following equation for $\mathbf{x}$:

$$\nabla_{\mathbf{w}} \hat{L}(f_{\mathbf{w}}) = 0 \tag{9.13}$$

By Eq. (9.11), we have that

$$\mathbf{X}^T \mathbf{X} \mathbf{w} - \mathbf{X}^T \mathbf{y} = 0 \tag{9.14}$$

That is,

$$\mathbf{w} = (\mathbf{X}^T \mathbf{X})^{-1} \mathbf{X}^T \mathbf{y} \tag{9.15}$$

Chapter 10
Deep Learning

This chapter is focused on a specific topic in modern machine learning, i.e., deep learning. First of all, we introduce a few fundamental aspects, including perceptron and why we need multi-layer structures, how the convolutional neural networks extract features layer by layer, the back-propagation learning algorithm, and the functional layers of convolutional neural networks. Then, we will focus on the safety and security vulnerabilities of deep learning, explaining uncertainty estimation, adversarial attack on the robustness, poisoning attack, model stealing, membership inference and model inversion. Unlike traditional machine learning models that we discussed in the previous chapters where the structure or the training algorithm of a machine learning model may be considered for the design of the attacks, most attacks for deep learning do not consider the internal structure of a deep learning model, although the gradient may be used in some cases. These black-box or grey-box attacks suggest that many of the attack algorithms for deep learning may also be applicable to traditional machine learning models, which explains why we frequently refer to algorithms in the chapter when discussing safety threads for traditional machine learning models.

10.1 Perceptron

10.1.1 Biological vs Artificial Neurons

A human brain contains billions of neurons, which—as shown in Fig. 10.1—are inter-connected nerve cells that are involved in processing and transmitting chemical and electrical signals. Neurons use dendrites as branches to receive information from other neurons. The received information is processed by the cell body or Soma. A neuron sends information to other neurons through a cable—called axon—and the synapse which connects an axon with other neurons' dendrites.

X. Huang et al., *Machine Learning Safety*, Artificial Intelligence: Foundations, Theory, and Algorithms, https://doi.org/10.1007/978-981-19-6814-3_10

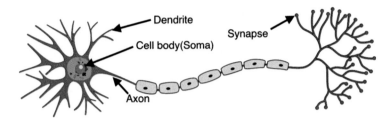

Fig. 10.1 Biological neuron (from Wikipedia)

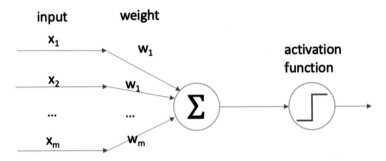

Fig. 10.2 Artificial neuron

Table 10.1 Mapping of concepts between biological and artificial neurons

Biological neuron	Artificial neuron
Dentrites	Input
Cell body (soma)	Node
Axon	Output
Synapse	Weights

In 1943, Warren McCullock and Walter Pitts published a simplified brain cell, called McCullock-Pitts (MCP) neuron. As shown in Fig. 10.2, an artificial neuron represents a nerve cell as a simple logic gate with binary outputs. It takes inputs, weighs them separately, sums them up, and passes this sum through a nonlinear function to produce output.

Table 10.1 is a brief conceptual mapping between biological and artificial neurons.

10.1.2 Learning Algorithm of Perceptron

A perceptron is an artificial neuron that does certain computations (such as detect features or execute business intelligence) on the input data. Perceptron was introduced by Frank Rosenblatt in 1957, when he proposed a perceptron learning rule based on the original MCP neuron. In July 1958, an IBM 704—a 5-ton computer with the size of a room—was fed a series of punch cards. After 50 trials, the computer taught itself to distinguish cards marked on the left from cards marked on the right [96]. It was a demonstration of the "perceptron", and was "the first machine which is capable of having an original idea," according to its creator, Frank Rosenblatt.

A perceptron learning algorithm is a supervised learning of binary classifiers. The binary classifier processes an input $\mathbf{x} = (x_1, \ldots, x_m)$ as follows:

1. Use one weight w_i per feature X_i;
2. Multiply weights w_i with the respective input features x_i of $\mathbf{x}$, and add bias w_0;
3. If the result is greater than a pre-specified threshold, return 1. Otherwise, return 0.

The weights $w_1, \ldots, w_m$ and bias w_0 of the binary classifier need to be learned. Given a set D of training instances, the learning algorithm proceeds as follows:

- Initialize weights randomly,
- Take one sample $(\mathbf{x}_i, y_i) \in D$ and make a prediction $\hat{y}_i$,
- For erroneous predictions, update weights with the following rules:

  - If the output is $\hat{y}_i = 0$ but the label is $y_i = 1$, increase the weights.
  - If the output is $\hat{y}_i = 1$ but the label is $y_i = 0$, decrease the weights.

  that is, we let

$$w_i = w_i + \Delta w_i \quad \text{such that} \quad \Delta w_i = \eta(y_i - \hat{y}_i)\mathbf{x}_i \qquad (10.1)$$

where $\eta \ll 1$ is a constant representing the learning rate.

10.1.3 Expressivity of Perceptron

While simple, perceptron is the foundation of modern deep learning, which uses a network of perceptrons where there are multiple layers and there are multiple neurons per layer. Why do we need a multi-layer perceptron (MLP) instead of just a single perceptron? This is a question related to the expressivity of a single perceptron. Actually, as we will show below, a single perceptron can express some useful functions but cannot do well for others.

10.1.4 Linearly Separable Function

As perceptron is a linear classifier, it is able to work well on all linearly separable datasets, i.e., classify instances that can be separated with a linear function.

Table 10.2 Truth table for
logic ∧ (And)

X_1	X_2	y
0	0	0
0	1	0
1	0	0
1	1	1

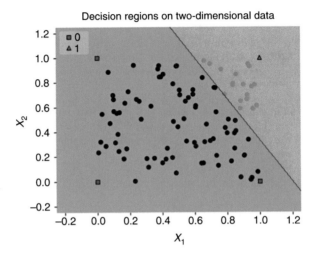

Fig. 10.3 Visualisation of a perceptron learned from the data instances in Table 10.2

Example 10.1 Table 10.2 presents an example dataset of four instances for the logic operator ∧. Actually, we generalise the logic operator ∧ to the following function.

$$f_\wedge(x_1, x_2) = \begin{cases} 1 & \text{if } x_1 > 0.5 \text{ and } x_2 > 0.5 \\ 0 & \text{otherwise} \end{cases} \tag{10.2}$$

In Fig. 10.3, we use squares to represent 0 and triangles to represent 1. By learning a perceptron, it is possible to get a linear separation function as exhibited in the figure. We use different colors to denote different areas in which the instances should be classified accordingly. In Fig. 10.3, we also generate a random sample of 100 instances and use the learned perceptron to predict the instances with different colors (with an accuracy close to 1.0).

Example 10.2 The other example is as shown in Table 10.3, presenting an example dataset of four instances for the logic operator ∨. Actually, we generalise the logic operator ∨ to the following function.

$$f_\vee(x_1, x_2) = \begin{cases} 1 & \text{if } x_1 > 0.5 \text{ or } x_2 > 0.5 \\ 0 & \text{otherwise} \end{cases} \tag{10.3}$$

Table 10.3 Truth table for
logic ∨ (Or)

X_1	X_2	y
0	0	0
0	1	1
1	0	1
1	1	1

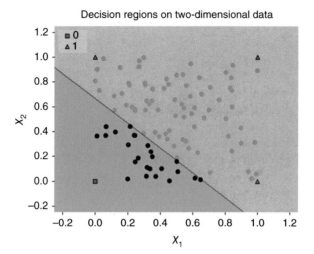

Fig. 10.4 Visualisation of a perceptron learned from the data instances in Table 10.3

Figure 10.4 presents the visualisation of the learned perceptron. We can see that, the separating line is different from that of Fig. 10.3. The separating lines in both figures are able to separate the data very well (with accuracy close to 1.0).

The above examples are all based on a 2-dimensional dataset. The learned perceptrons are actually a line on the 2-dimensional space. When the dataset is d-dimensional, the perceptron is a d-dimensional hyper-plane.

10.1.5 Linearly Inseparable Function

Unfortunately, not all datasets are linearly separable.

Example 10.3 Table 10.4 is a dataset of four instances, which is generated from the following function:

$$f_\oplus(x_1, x_2) = \begin{cases} 1 & \text{if } x_1 > 0.5 \text{ and } x_2 < 0.5 \\ 1 & \text{if } x_1 < 0.5 \text{ and } x_2 > 0.5 \\ 0 & \text{otherwise} \end{cases} \tag{10.4}$$

Table 10.4 Truth table for logic ⊕ (XOR)

X_1	X_2	y
0	0	0
0	1	1
1	0	1
1	1	0

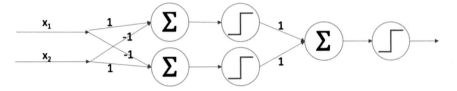

Fig. 10.5 A two-layer perceptron to solve XOR problem

While we can still apply perceptron, the learned perceptron cannot reach high accuracy (~ 0.5 accuracy).

Therefore, this example shows that the perceptron in itself does not have sufficient expressiveness to work with complex functions. This contributes as the reason why we have to consider multiple layers and more neurons.

10.1.6 Multi-Layer Perceptron

To deal with the XOR problem, a two-layer perceptron as shown in Fig. 10.5 has been suggested.

where the activation function is $ReLU(x) = \max(0, x)$.

If written in the matrix form, we have the following expressions for the training data, which shows that the above two-layer perceptron perfectly classifies the training data.

$$\begin{pmatrix} 0 & 0 \\ 0 & 1 \\ 1 & 0 \\ 1 & 1 \end{pmatrix} \times \begin{pmatrix} 1 & 1 \\ -1 & 1 \end{pmatrix} = \begin{pmatrix} 0 & 0 \\ -1 & -1 \\ 1 & -1 \\ 0 & 0 \end{pmatrix} \text{ and } ReLU\begin{pmatrix} 0 & 0 \\ -1 & -1 \\ 1 & -1 \\ 0 & 0 \end{pmatrix} = \begin{pmatrix} 0 & 0 \\ 0 & 1 \\ 1 & 0 \\ 0 & 0 \end{pmatrix}$$

$$\tag{10.5}$$

$$\begin{pmatrix} 0 & 0 \\ 0 & 1 \\ 1 & 0 \\ 0 & 0 \end{pmatrix} \times \begin{pmatrix} 1 \\ 1 \end{pmatrix} = \begin{pmatrix} 0 \\ 1 \\ 1 \\ 0 \end{pmatrix} \text{ and } ReLU\begin{pmatrix} 0 \\ 1 \\ 1 \\ 0 \end{pmatrix} = \begin{pmatrix} 0 \\ 1 \\ 1 \\ 0 \end{pmatrix} \tag{10.6}$$

10.1.7 *Practice*

10.1.7.1 Train a Perceptron

First of all, we load and prepare datasets.

```
from sklearn import datasets
dataset = datasets.load_digits()
X = dataset.data
y = dataset.target

observations = len(X)
features = len(dataset.feature_names)
classes = len(dataset.target_names)
print("Number of Observations: " + str(observations))
print("Number of Features: " + str(features))
print("Number of Classes: " + str(classes))

from sklearn.model_selection import train_test_split
X_train, X_test, y_train, y_test = train_test_split(X, y,
    test_size=0.20)
```

Then, we can call **sklearn**'s library function to train a perceptron model.

```
from sklearn.linear_model import Perceptron

clf = Perceptron(tol=1e-3, random_state=0)
clf.fit(X_train, y_train)
print("Training accuracy is %s"% clf.score(X_train,y_train))
print("Test accuracy is %s"% clf.score(X_test,y_test))

print("Labels of all instances:\n%s"%y_test)
y_pred = clf.predict(X_test)
print("Predictive outputs of all instances:\n%s"%y_pred)

from sklearn.metrics import classification_report,
    confusion_matrix
print("Confusion Matrix:\n%s"%confusion_matrix(y_test, y_pred))
print("Classification Report:\n%s"%classification_report(y_test,
    y_pred))
```

10.1.7.2 Display Class Regions for Boolean Functions

In the following, we present how to generate the visualisation as in Figs. 10.3 and 10.4. First of all, we install a package **mlxtend**.

```
$ pip3 install mlxtend
```

Then, we need to load the data instances $\mathbf{X} = \{(0, 0), (0, 1), (1, 0), (1, 1)\}$ and their labels. Note that, in the below code, we use **logical_and**. You are able to use others such as **logical_or** and **logical_xor**.

```
import numpy as np
# Loading data
X_train = np.array([[0.0,0.0],[0.0,1.0],[1.0,0.0],[1.0,1.0]])
y_train = np.array(np.logical_and(X_train[:, 0] > 0.5, X_train[:,
    1] > 0.5),
                dtype=int)
```

Once the data is loaded, we train a Perceptron, with initial parameters **w** = (1.5, 1.5). Note that, this is simply for the experiment, and it can be initialised to other values, or set as default by ignoring the parameter **coef_init**.

```
# Training a classifier
from sklearn.linear_model import Perceptron
clf = Perceptron(tol=1e-3, random_state=0)
clf.fit(X_train, y_train, coef_init=np.array([[1.5],[1.5]]))
```

We can print the final learned weights.

```
print(clf.coef_)
```

Finally, we can plot the regions for the classes.

```
# Plotting decision regions
from mlxtend.plotting import plot_decision_regions
plot_decision_regions(X_train, y_train, clf=clf, legend=2)
```

We may also generate a set of random points and plot them.

```
# Plotting randomly generated points
import matplotlib
import matplotlib.pyplot as plt
import random

n_sample = 100
X_test = np.array([[random.random() for i in range(2)] for j in
    range(n_sample)])
y_test = np.array(np.logical_and(X_test[:,0]>0.5,X_test[:,1]>0.5)
    ,dtype=int)
y_pred = clf.predict(X_test)
print(clf.score(X_test,y_test))
colors = matplotlib.cm.rainbow(np.linspace(0, 1, 5))
plt.scatter(X_test[:, 0],X_test[:, 1],color=[colors[i] for i in
    y_pred])
```

```
# Adding axes annotations
import matplotlib.pyplot as plt
plt.xlabel('X_1')
plt.ylabel('X_2')
plt.title('Decision regions on two-dimensional data')
plt.show()
```

10.2 Functional View

A deep neural network can be seen as a family of parametric, non-linear, and hierarchical representation learning functions. These learning functions are massively optimized with stochastic gradient descent (SGD) over some pre-specified objectives, such as the loss of a set of training instances. It is expected that, the learned functions will encode domain knowledge that is implicitly presented in the training instances.

10.2.1 Mappings Between High-Dimensional Spaces

Consider a feed-forward network as shown in Fig. 10.6. Assume that it has $m + 1$ layers, where Layer-0 is the input layer, Layer-m is the output layer, and Layer-1 to Layer-$(m - 1)$ are the hidden layers.

Every layer is a function, so we have functions $f_1, \ldots, f_m$, for hidden layers and output layer, and because every function is parameterised, we use $\mathbf{W} = \{\mathbf{W}_1, \ldots, \mathbf{W}_m\}$ to denote their parameters. Based on these, a neural network can be written in a functional way as follows.

$$f_{\mathbf{W}}(\mathbf{x}; \mathbf{W}_1, \ldots, \mathbf{W}_m) = f_m(f_{m-1}(\ldots f_1(\mathbf{x}; \mathbf{W}_1), \mathbf{W}_{m-1}); \mathbf{W}_m) \qquad (10.7)$$

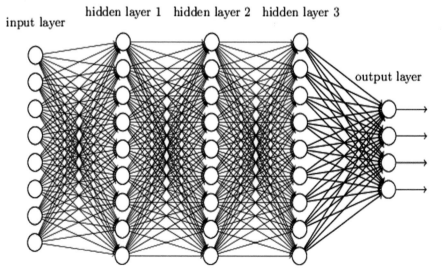

Fig. 10.6 A 5 layer feed-forward network

Alternatively, we may write

$$
\begin{aligned}
f_{\mathbf{W}}(\mathbf{x}) = \mathbf{v}_m \ \ &= f_m(\mathbf{v}_{m-1}; \mathbf{W}_m) \\
\mathbf{v}_{m-1} &= f_{m-1}(\mathbf{v}_{m-2}; \mathbf{W}_{m-1}) \\
&\dots \\
\mathbf{v}_2 \ \ &= f_2(\mathbf{v}_1; \mathbf{W}_2) \\
\mathbf{v}_1 \ \ &= f_1(\mathbf{v}_0; \mathbf{W}_1)
\end{aligned}
\tag{10.8}
$$

where $\mathbf{v}_i$ is the output value of Layer-i. Note that, we have both $\mathbf{v}_i$ and $\mathbf{W}_i$ as vectors or matrices, because it is typical that there are many neurons per layer and the layer functions are parameterised with many parameters.

Take a closer look at Eq. (10.8), given a function $\mathbf{v}_i = f_i(\mathbf{v}_{i-1}; \mathbf{W}_i)$, once the parameters $\mathbf{W}_i$ are learned, it is a transformation from $\mathbf{v}_{i-1}$ to $\mathbf{v}_i$. Let each layer-i have k_i neurons, we have that $\mathbf{v}_{i-1}$ is a vector of k_{i-1} entries and $\mathbf{v}_i$ is a vector of k_i entries. Therefore, the transformation can be seen as a mapping from high-dimensional space $\mathbb{R}^{k_{i-1}}$ to $\mathbb{R}^{k_i}$. Generalise this to the entire network, we have

$$
\mathbb{R}^{k_0} \xrightarrow{f_1} \mathbb{R}^{k_1} \xrightarrow{f_2} \dots \xrightarrow{f_m} \mathbb{R}^{k_m}
\tag{10.9}
$$

Note that, k_0 is the number of input features and k_m is the number of class labels.

10.2.2 Training Objective

The training typically intends to get the best weights $\mathbf{W}^*$ as follows.

$$
\mathbf{W}^* \leftarrow \arg\min_{\mathbf{W}} \sum_{(\mathbf{x}, y) \in D} L(y, \mathbf{v}_m)
\tag{10.10}
$$

where $L(y, f_{\mathbf{W}}(\mathbf{x}))$ is typically a loss function measuring the gap between actual label y with its current prediction $f_{\mathbf{W}}(\mathbf{x})$. However, the optimisation problem is highly dimensional and non-convex. Therefore, in most cases, the training ends up with an approximation $\hat{\mathbf{W}}$.

10.2.3 Recurrent Neural Networks

The above is mainly for feedforward neural networks (FNNs), which model a function $\phi : X \to Y$ that maps from input domain X to output domain Y: given an input $x \in X$, it outputs the prediction $y \in Y$. For a sequence of inputs $x_1, \dots, x_n$, an FNN ϕ considers each input individually, that is, $\phi(x_i)$ is independent from $\phi(x_{i+1})$.

By contrast, a recurrent neural network (RNN) processes an input sequence by iteratively taking inputs one by one. A recurrent layer can be modeled as a function $\psi : X' \times C \times Y' \to C \times Y'$ such that $\psi(x_t, c_{t-1}, h_{t-1}) = (c_t, h_t)$ for $t = 1, \ldots, n$, where t denotes the t-th time step, c_t is the cell state used to represent the intermediate memory and h_t is the output of the t-th time step. More specifically, the recurrent layer takes three inputs: x_t at the current time step, the prior memory state c_{t-1} and the prior cell output h_{t-1}; consequently, it updates the current cell state c_t and outputs current h_t.

RNNs differ from each other given their respective definitions, i.e., internal structures, of recurrent layer function ψ, of which long short-term memory (LSTM) in Eq. (10.11) is the most popular and commonly used one.

$$
\begin{aligned}
f_t &= \sigma(W_f \cdot [h_{t-1}, x_t] + b_f) \\
i_t &= \sigma(W_i \cdot [h_{t-1}, x_t] + b_i) \\
c_t &= f_t * c_{t-1} + i_t * \tanh(W_c \cdot [h_{t-1}, x_t] + b_c) \\
o_t &= \sigma(W_o \cdot [h_{t-1}, x_t] + b_o) \\
h_t &= o_t * \tanh(c_t)
\end{aligned}
\tag{10.11}
$$

such that $\sigma(x) \in [0, 1]$ for any $x \in \mathbb{R}$, tanh is the hyperbolic tangent function such that $\tanh(x) \in [-1, 1]$ for any $x \in \mathbb{R}$, W_f, W_i, W_c, W_o are weight matrices, b_f, b_i, b_c, b_o are bias vectors, f_t, i_t, o_t are internal gate variables, h_t is the hidden state variable (utilising o_t), and c_t is the cell state variable. For the connection with successive layers, we only take the last output h_n as the output. For simplicity, when working with finite sequential data, we can also define a recurrent layer as $\psi :$ $(X')^n \to Y'$, which takes, as input, a sequential data of length n and returns the last output h_n. Figure 10.7 presents an illustrative diagram for LSTM cell.

In LSTM, σ is the sigmoid function and tanh is the hyperbolic tangent function; W and b represent the weight matrix and bias vector, respectively; f_t, i_t, o_t are internal gate variables of the cell. In general, the recurrent layer (or LSTM layer)

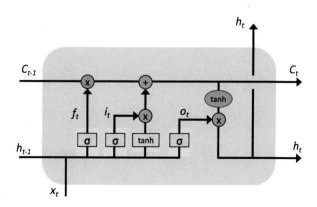

Fig. 10.7 LSTM cell

is connected to non-recurrent layers such as fully connected layers so that the cell output propagates further. We denote the remaining layers with a function $\phi_2 : Y' \rightarrow Y$. Meanwhile, there can be feedforward layers connecting to the RNN layer, and we let it be another function $\phi_1 : X \rightarrow X'$. As a result, the RNN model that accepts a sequence of inputs $x_1, \ldots, x_n$ can be modeled as a function φ such that $\varphi(x_1 \ldots x_n) = \phi_2 \cdot \psi(\prod_{i=1}^{n} \phi_1(x_i))$. Normally, the recurrent layer is connected to non-RNN layers such as fully connected layers so that the output h_n is processed further. We let the remaining layer be a function $\phi_2 : Y' \rightarrow Y$. Moreover, there can be feedforward layers connecting to the RNN layer, and we let it be a function $\phi_1 : X \rightarrow X'$. Then given a sequential input $x_1, \ldots, x_n$, the RNN is a function ϕ such that the features of the sequential input are extracted by ϕ_1 before being processing by the recurrent layer ψ and classified with ϕ_2.

10.2.4 Learning Representation and Features

10.2.4.1 Raw Digital Representation

Every instance has to be represented in a digital form. For example, the 8th instance in the **digits** dataset is an image of digit 8 as shown in Fig. 10.8.

Actually, it is stored as a matrix as follows:

$$
\begin{pmatrix}
0 & 0 & 9 & 14 & 8 & 1 & 0 & 0 \\
0 & 0 & 12 & 14 & 14 & 12 & 0 & 0 \\
0 & 0 & 9 & 10 & 0 & 15 & 4 & 0 \\
0 & 0 & 3 & 16 & 12 & 14 & 2 & 0 \\
0 & 0 & 4 & 16 & 16 & 2 & 0 & 0 \\
0 & 3 & 16 & 8 & 10 & 13 & 2 & 0 \\
0 & 1 & 15 & 1 & 3 & 16 & 8 & 0 \\
0 & 0 & 11 & 16 & 15 & 11 & 1 & 0
\end{pmatrix}
\tag{10.12}
$$

Fig. 10.8 A small image of digit 8

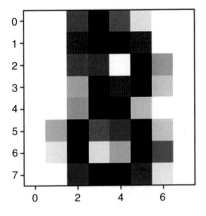

As another example, the videos are actually a sequence of images, and therefore it is stored as a 3-dimensional array.

Features are domain dependent. Evidently, for computer vision, pixels are input features, and for natural language processing, words are input features. In addition to input features which closely relate to data representation, we may use the term latent features or hidden features for those features in the hidden layers.

10.2.4.2 Feature Extraction

Feature extraction is one of the key intermediate tasks for learning, for both traditional machine learning and deep learning. It is a process that identifies important features or attributes of the data. For traditional machine learning, as shown in Fig. 10.9, it first extracts features and then applies a learnable classifier. The feature extraction is treated as a step independent of the classification. There are many different methods for feature extraction, for example, SIFT (scale-invariant feature transform).

Deep learning, however, requires only one step (i.e., end-to-end) to implement both feature extraction and classification, as shown in Fig. 10.10. Both the feature extractor and the classifier are trained at the same time.

Feature extraction is closely related to dimensionality reduction, i.e., to separate data as much as possible. Most data distributions and tasks are non-linear, so a linear assumption is often convenient, but not necessarily truthful. Therefore, to get non-linear machines without too much effort, we may have to consider non-linear features.

There are many ways to get non-linear features, including e.g.,

- application of non-linear kernels, e.g., polynomial, RBF, etc.
- explicit design of features, e.g., SIFT, HOG, etc.

Fig. 10.9 Flow of traditional machine learning

Fig. 10.10 Flow of deep learning

The quality of features is usually evaluated against the following few criteria:

- invariance
- repeatabilty
- discriminativeness
- robustness

It is useful to note that these criteria may be conflicting. Hence, there needs to be a trade-off between criteria.

10.2.4.3 Data Manifold

Actually, most natural, high-dimensional data (e.g. faces) lie on lower dimensional manifolds. For example, Fig. 10.11 is the so-called "swiss roll", where the data points are 3-dimensional, but they all lie on a 2-dimensional manifold. That is, the actual dimensionality of the manifold is 2, while the dimensionality of the input space is 3.

Therefore, although the data points may consist of thousands of features, they can be described as a function of only a few underlying parameters. That is, the data points are actually sampled from a low-dimensional manifold that is embedded in a high-dimensional space.

10.2.4.4 Difficulties of Simply Using Dimensionality Reduction or Kernel

The above observation suggests that our goal should be on discovering lower dimensional manifolds. We remark that, these manifolds are most probably highly non-linear.

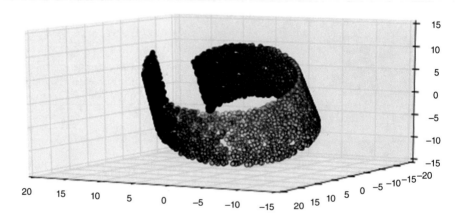

Fig. 10.11 Data manifold—"swiss role" example

The success of this requires two hypotheses:

- If we can compute the coordinates of the input (e.g., a face image) to this non-linear manifold then the data become separable. This hypothesis suggests the existence of *functional mapping*. For the "swiss role" example, there should be a (non-linear) function mapping from 3d space to 2d space, on which the data can be linearly separable.
- Semantically similar things lie closer than semantically dissimilar things. This implies the existence of applicable dimensional reduction methods.

While raw data live in huge dimensionality, semantically meaningful raw data prefer lower dimensional manifolds, which still live in the same huge dimensionality. Can we discover this manifold to embed our data on?

10.2.4.5 End-to-End Learning of Feature Hierarchies

The above discussions basically suggest that, it is an almost impossible task to manually craft features and also nontrivial to design algorithms (dimensionality reduction, functional mapping, etc) to compute features. This is in stark contrast with deep learning. Actually, one of the key advantages of convolutional neural networks is their ability to learn (or extract) features automatically.

In a CNN, there are a pipeline of successive layers, such that each layer's output is the input for the next layer. Layers produce features of higher and higher abstractions, such that the shadow layers extract low-level features (e.g. edges or corners), middle layers extract mid-level features (e.g. circles, squares, textures), and deep layers capture high level, class-specific features (e.g. face detector). See Fig. 10.12.

We remark that, for CNNs, it has been shown that, preferably, training data should be as raw as possible. That is, no additional feature extraction phase is needed.

10.2.4.6 Why Learn the Features?

Manually designed features often take a lot of time to come up with and implement, a lot of time to validate, and are incomplete, as one cannot know if they are optimal for the task. On the other hand, learned features are easy to adapt, very compact and specific to the task at hand. Given a basic architecture in mind, it is relatively easy and fast to optimize, i.e., time spent on designing features is now spent on designing architectures.

10.2.5 Practice

The following is a code to train a neural network with fully-connected layers

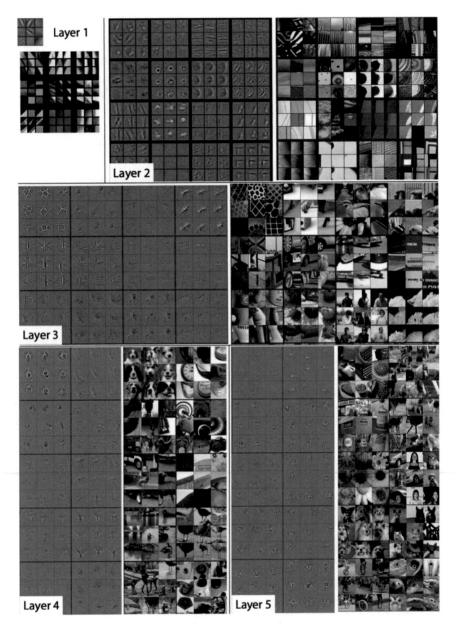

Fig. 10.12 Visualisation of features in hidden layers [191]

10.2.5.1 Train a Fully Connected Model

First of all, we install two packages torch and torchvision.

```
$ pip3 install torch
$ pip3 install torchvision
```

Then, we set up hyper-parameters (e.g., batchsize, epoch, learning rate), device (e.g., CPU or GPU), and load training dataset (MNIST).

```python
import torch
import torch.nn as nn
import torch.nn.functional as F
import torch.optim as optim
from torchvision import datasets, transforms
import argparse
import time

# Setup hyper-parameter
parser = argparse.ArgumentParser(description='PyTorch MNIST
    Training')
parser.add_argument('--batch-size', type=int, default=128,
    metavar='N',
                    help='input batch size for training (default:
    128)')
parser.add_argument('--test-batch-size', type=int, default=128,
    metavar='N',
                    help='input batch size for testing (default:
    128)')
parser.add_argument('--epochs', type=int, default=10, metavar='N'
    ,
                    help='number of epochs to train')
parser.add_argument('--lr', type=float, default=0.01, metavar='LR
    ',
                    help='learning rate')
parser.add_argument('--no-cuda', action='store_true', default=
    False,
                    help='disables CUDA training')
parser.add_argument('--seed', type=int, default=1, metavar='S',
                    help='random seed (default: 1)')

args = parser.parse_args(args=[])

# Judge cuda is available or not
use_cuda = not args.no_cuda and torch.cuda.is_available()
#device = torch.device("cuda" if use_cuda else "cpu")
device = torch.device("cpu")

torch.manual_seed(args.seed)
kwargs = {'num_workers': 1, 'pin_memory': True} if use_cuda else
    {}

# Setup data loader
```

```
35  transform=transforms.Compose([
36          transforms.ToTensor(),
37          transforms.Normalize((0.1307,), (0.3081,))
38          ])
39  trainset = datasets.MNIST('../data', train=True, download=True,
40                      transform=transform)
41  testset = datasets.MNIST('../data', train=False,
42                      transform=transform)
43  train_loader = torch.utils.data.DataLoader(trainset,batch_size=
        args.batch_size, shuffle=True,**kwargs)
44  test_loader = torch.utils.data.DataLoader(testset,batch_size=args
        .test_batch_size, shuffle=False, **kwargs)
```

We can define a fully connected network as follows, with the structure 784-128-64-32-10,

```
1   # Define fully connected network
2   class Net(nn.Module):
3       def __init__(self):
4           super(Net, self).__init__()
5           self.fc1 = nn.Linear(28*28, 128)
6           self.fc2 = nn.Linear(128, 64)
7           self.fc3 = nn.Linear(64, 32)
8           self.fc4 = nn.Linear(32, 10)
9
10      def forward(self, x):
11          x = self.fc1(x)
12          x = F.relu(x)
13          x = self.fc2(x)
14          x = F.relu(x)
15          x = self.fc3(x)
16          x = F.relu(x)
17          x = self.fc4(x)
18          output = F.log_softmax(x, dim=1)
19          return output
```

Then, we define the training function, which computes loss and updates parameters for each minibatch.

```
1   # Training function
2   def train(args, model, device, train_loader, optimizer, epoch):
3       model.train()
4       for batch_idx, (data, target) in enumerate(train_loader):
5           data, target = data.to(device), target.to(device)
6           data = data.view(data.size(0),28*28)
7
8           # Clear gradients
9           optimizer.zero_grad()
10
11          # Compute loss
12          loss = F.cross_entropy(model(data), target)
13
14          # Get gradients and update
15          loss.backward()
16          optimizer.step()
```

We also can define a predict function, which outputs training loss and test loss for each epoch.

```
# Predict function
def eval_test(model, device, test_loader):
    model.eval()
    test_loss = 0
    correct = 0
    with torch.no_grad():
        for data, target in test_loader:
            data, target = data.to(device), target.to(device)
            data = data.view(data.size(0),28*28)
            output = model(data)
            test_loss += F.cross_entropy(output, target,
size_average=False).item()
            pred = output.max(1, keepdim=True)[1]
            correct += pred.eq(target.view_as(pred)).sum().item()
    test_loss /= len(test_loader.dataset)
    test_accuracy = correct / len(test_loader.dataset)
    return test_loss, test_accuracy
```

Finally, we define the main function and call the training function for each epoch.

```
# Main function, train the dataset and print training loss, test
    loss
def main():
    model = Net().to(device)
    optimizer = optim.SGD(model.parameters(), lr=args.lr)
    for epoch in range(1, args.epochs + 1):
        start_time = time.time()

        # Training
        train(args, model, device, train_loader, optimizer, epoch
)

        # Get trnloss and testloss
        trnloss, trnacc = eval_test(model, device, train_loader)
        tstloss, tstacc = eval_test(model, device, test_loader)

        # Print trnloss and testloss
        print('Epoch '+str(epoch)+': '+str(int(time.time()-
start_time))+'s', end=', ')
        print('trn_loss: {:.4f}, trn_acc: {:.2f}%'.format(trnloss
, 100. * trnacc), end=', ')
        print('test_loss: {:.4f}, test_acc: {:.2f}%'.format(
tstloss, 100. * tstacc))

if __name__ == '__main__':
    main()
```

10.3   Forward and Backward Computation

In Sect. 10.1, we have explained that a multi-layer perceptron has more expressive power than a single-layer perceptron. In particular, it is able to find a two-layer perceptron to solve the XOR problem, while it is not possible for a single-layer perceptron. However, we did not explain in Sect. 10.1 how to compute the weights for the two-layer perceptron for XOR. Moreover, we note that, the perceptron learning algorithm cannot be used for learning multi-layer perceptron. Actually, the learning algorithm for multi-layer perceptron, called backpropagation (BP), is one of the key milestones for the development of deep learning.

The BP algorithm computes the gradient of the loss function with respect to each weight by the chain rule. Instead of computing one gradient for each weight, it is able to compute the gradient for one layer at a time, and more importantly, it is able to iterate backwards from the last layer to avoid redundant calculations of intermediate terms in the chain rule. This makes it efficient enough to train large scale neural networks.

The BP algorithm is the foundation of deep learning, and in this section, we use a running example to explain its computation.

10.3.1 Running Example

In this example, we have three layers, one input layer, one hidden layer, and one output layer. Each layer has two neurons. The connections between neurons are given in the left diagram of Fig. 10.13.

The diagram on the right is an illustration of a neuron with an activation function. Each neuron has two values u and v, representing its values before and after the application of the activation function, respectively.

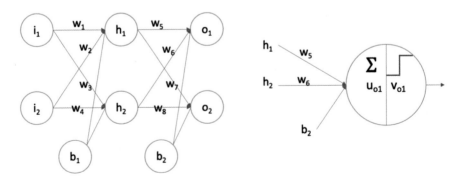

Fig. 10.13 A simple neural network with a hidden layer (Left) and the illustration of a neuron with activation function (Right)

The learning is a dynamic process with weights continuously updated until convergence. Now, we assume that, at some point, the current weights are

$$\begin{pmatrix} w_1 & w_2 \\ w_3 & w_4 \end{pmatrix} = \begin{pmatrix} 0.15 & 0.20 \\ 0.25 & 0.30 \end{pmatrix}, \begin{pmatrix} w_5 & w_6 \\ w_7 & w_8 \end{pmatrix} = \begin{pmatrix} 0.40 & 0.45 \\ 0.50 & 0.55 \end{pmatrix}, \begin{pmatrix} b_1 \\ b_2 \end{pmatrix}$$

$$= \begin{pmatrix} 0.35 \\ 0.60 \end{pmatrix} \tag{10.13}$$

Note that, every row in a weight matrix stores the input weights for a neuron, for example, the row $(0.15\ 0.20)$ is associated with the neuron h_1. Also, each entry in a bias vector represents a bias for a layer.

10.3.2 Forward Computation

The first step of each iteration of the BP algorithm is to compute the loss by making a forward computation. Assume that we have an input $\mathbf{x} = (0.05, 0.10)^T$ for the network in Fig. 10.13, we can have

$$\begin{pmatrix} u_{h_1} \\ u_{h_2} \end{pmatrix} = \begin{pmatrix} 0.15 & 0.20 \\ 0.25 & 0.30 \end{pmatrix} \times \begin{pmatrix} 0.05 \\ 0.10 \end{pmatrix} + \begin{pmatrix} 0.35 \\ 0.35 \end{pmatrix} = \begin{pmatrix} 0.3775 \\ 0.6425 \end{pmatrix} \tag{10.14}$$

Assume that the network uses the Sigmoid function σ as the activation function, we have

$$\begin{pmatrix} v_{h_1} \\ v_{h_2} \end{pmatrix} = \sigma\left(\begin{pmatrix} 0.3775 \\ 0.3925 \end{pmatrix}\right) \approx \begin{pmatrix} 0.5927 \\ 0.5969 \end{pmatrix} \tag{10.15}$$

as the output of the hidden layer. On the output layer, we have

$$\begin{pmatrix} u_{o_1} \\ u_{o_2} \end{pmatrix} = \begin{pmatrix} 0.40 & 0.45 \\ 0.50 & 0.55 \end{pmatrix} \times \begin{pmatrix} 0.5927 \\ 0.5969 \end{pmatrix} + \begin{pmatrix} 0.60 \\ 0.60 \end{pmatrix} = \begin{pmatrix} 1.1057 \\ 1.2247 \end{pmatrix} \tag{10.16}$$

Consider the Sigmoid function, we have

$$\begin{pmatrix} v_{o_1} \\ v_{o_2} \end{pmatrix} = \sigma\left(\begin{pmatrix} 1.1057 \\ 1.2247 \end{pmatrix}\right) \approx \begin{pmatrix} 0.7513 \\ 0.7729 \end{pmatrix} \tag{10.17}$$

as the output of the output layer. That is, $\hat{y} = (0.7513, 0.7729)$. Now, assuming that the label of $\mathbf{x}$ is $y = (0.01, 0.99)^T$ and we are using the mean square error, we can compute the loss for

$$L(\mathbf{x}, y) = \frac{1}{2}(0.7513 - 0.01)^2 + \frac{1}{2}(0.7729 - 0.99)^2 \approx 0.2748 + 0.0236 = 0.2984 \tag{10.18}$$

where we let $L_{o1}(\mathbf{x}, y) = \frac{1}{2}(0.7513 - 0.01)^2$ and $L_{o2}(\mathbf{x}, y) = \frac{1}{2}(0.7729 - 0.99)^2$, representing the loss of individual neurons o_1 and o_2, respectively.

10.3.3   Backward Computation

Once we have the loss $L(\mathbf{x}, y)$, we can start back-propagation by applying the chain rule.

10.3.3.1   Weights of Output Neurons

First of all, for the weights of output neuron, such as w_5, we can compute as follows:

$$\frac{\partial L}{\partial w_5} = \frac{\partial L_{o1}}{\partial v_{o1}} * \frac{\partial v_{o1}}{\partial u_{o1}} * \frac{\partial u_{o1}}{\partial w_5} \tag{10.19}$$

Figure 10.14 presents an illustrative diagram for the Eq. (10.19). Actually, the backpropagation goes from the loss L_{o1} to the value v_{o1}, u_{o1}, until the weight w_5.

Concretely, for the running example, we have

$$\frac{\partial L_{o1}}{\partial v_{o1}} = \frac{\partial}{\partial v_{o1}} (\frac{1}{2}(y^{(1)} - v_{o1})^2) = -(y^{(1)} - v_{o1}) = 0.7513 - 0.01 = 0.74 \tag{10.20}$$

Fig. 10.14 Backward propagation on the output neuron

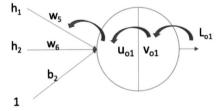

where $y^{(1)}$ is the first component of y, and

$$\frac{\partial v_{o_1}}{\partial u_{o_1}} = \frac{\partial \sigma(u_{o_1})}{\partial u_{o_1}} = \sigma(u_{o_1})(1 - \sigma(u_{o_1})) = v_{o_1}(1 - v_{o_1}) \approx 0.7513$$
$$\times 0.2487 \approx 0.1868 \tag{10.21}$$

and

$$\frac{\partial u_{o_1}}{\partial w_5} = \frac{\partial}{\partial w_5}(w_5 v_{h_1} + w_6 v_{h_2} + b_2) = v_{h_1} \approx 0.5927 \tag{10.22}$$

Therefore, we have

$$\frac{\partial L}{\partial w_5} = \frac{\partial L_{o_1}}{\partial v_{o_1}} * \frac{\partial v_{o_1}}{\partial u_{o_1}} * \frac{\partial u_{o_1}}{\partial w_5} \approx 0.74 \times 0.1868 \times 0.5927 = 0.0819 \tag{10.23}$$

10.3.3.2   Weights of Hidden Neurons

Now, the weight of hidden layer can be done recursively by applying the chain rules, e.g.,

$$\begin{aligned}
\frac{\partial L}{\partial w_1} &= \frac{\partial L}{\partial v_{h_1}} * \frac{\partial v_{h_1}}{\partial u_{h_1}} * \frac{\partial u_{h_1}}{\partial w_1} \\
&= (\frac{\partial L_{o_1}}{\partial v_{h_1}} + \frac{\partial L_{o_2}}{\partial v_{h_1}}) * \frac{\partial v_{h_1}}{\partial u_{h_1}} * \frac{\partial u_{h_1}}{\partial w_1} \\
&= (\frac{\partial L_{o_1}}{\partial u_{o_1}}\frac{\partial u_{o_1}}{\partial u_{h_1}} + \frac{\partial L_{o_2}}{\partial u_{o_2}}\frac{\partial u_{o_2}}{\partial u_{h_1}}) * \frac{\partial v_{h_1}}{\partial u_{h_1}} * \frac{\partial u_{h_1}}{\partial w_1} \\
&= (\frac{\partial L_{o_1}}{\partial u_{o_1}}w_5 + \frac{\partial L_{o_2}}{\partial u_{o_2}}w_7) * \frac{\partial v_{h_1}}{\partial u_{h_1}} * \frac{\partial u_{h_1}}{\partial w_1}
\end{aligned} \tag{10.24}$$

Note that, all the components of Eq. (10.24) can now be computed as the method we used for the output layer. Figure 10.15 presents an illustration of backward propagation as in Eq. (10.24).

10.3.3.3   Weight Update

Finally, once we compute the gradients $\frac{\partial L}{\partial w_5}$ or $\frac{\partial L}{\partial w_1}$, we can update the weights w_5 or w_1 by applying the gradient descent algorithm. We remark that, while the above computation is conducted for individual weights, the BP algorithm can work on a layer basis to significantly improve the efficiency.

Fig. 10.15 Backward
propagation on the hidden
neuron

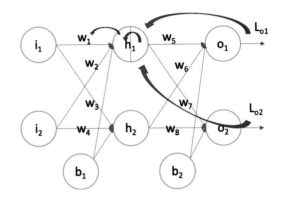

10.3.4   Regularisation as Constraints

In general, regularisation is a set of methods to prevent overfitting or help the
optimization. Typically, this is done by having additional terms in the training
optimisation objective.

10.3.4.1   Overfitting

Overfitting is a concept closely related to the generalisation error as introduced
in Sect. 3.1, which is the gap between empirical loss and expected loss. It has
been observed that, the following two reasons may contribute as the key to the
overfitting.

- dataset is too small
- hypothesis space is too large

To understand the second point, we note that, the larger the hypothesis space, the
easier is for a learning algorithm to find a hypothesis that has small training error.
However, finding a small training error does not warrant the found hypothesis can
be of a small test error, and so this may lead to large test error (overfitting). This
observation suggests that it might be beneficial to leave out useless hypotheses,
which is what regularization is for.

10.3.4.2   Regularization as Hard Constraint

Assume that, for a dataset $D = (\mathbf{X}, \mathbf{y})$ of n training instances, we have the following
optimising objective:

$$\min_{f} L(f, D) = \frac{1}{n} \sum_{i=1}^{n} L(f, \mathbf{x}_i, y_i) \tag{10.25}$$

subject to $f \in \mathcal{H}$

Considering that for a deep learning model, f is parameterised over the weights W, we have

$$\min_{W} L(W, D) = \frac{1}{n} \sum_{i=1}^{n} L(W, \mathbf{x}_i, y_i)$$

$$\text{subject to} \quad W \in \mathbb{R}^{|W|}$$

$$(10.26)$$

where $|W|$ is the number of weights.

The regularisation is to add further constraints. For example, if we ask for L_2 regularisation, we have

$$\min_{W} L(W, D) = \frac{1}{n} \sum_{i=1}^{n} L(W, \mathbf{x}_i, y_i)$$

$$\text{subject to} \quad W \in \mathbb{R}^{|W|}$$

$$||W||_2^2 \leq r^2$$

$$(10.27)$$

for some pre-specified $r > 0$.

10.3.4.3   Regularization as Soft Constraint

While the hard constraints limit the selection of hypothesis, it might not be easy to be integrated with the backpropagation algorithm, which does not consider the constraints directly. This can be done through a soft constraint, e.g.,

$$\min_{W} L(W, D) = \frac{1}{n} \sum_{i=1}^{n} L(W, \mathbf{x}_i, y_i) + \lambda ||W||_2^2$$

$$(10.28)$$

where λ is a hyper-parameter to balance between the loss term and the constraint/penalty term. Alternatively, this can be done through Lagrangian multiplier method:

$$\min_{W} L(W, D) = \frac{1}{n} \sum_{i=1}^{n} L(W, \mathbf{x}_i, y_i) + \lambda (||W||_2^2 - r)$$

$$(10.29)$$

10.3.5   Practice

The following code is to save the weights of a trained model and load the weights from a file.

```
# Save model
torch.save(model.state_dict(), 'model.pt')

# Load model
model.load_state_dict(torch.load('model.pt'))
```

10.4   Convolutional Neural Networks

This section introduces a specific class of neural networks that has been shown very effective in processing images. In 1998, Yann LeCun and his collaborators developed a neural network for handwritten digits called LeNet [94]. It is a feedforward network with several hidden layers, trained with the backpropagation algorithm. It is later formalised with the name convolutional neural networks (CNNs). Since LeNet, there are many other variants of CNNs, such as AlexNet, VGG16, ResNet, GoogLeNet, and so on. These variants introduce new functional layers or training methods that help improve the performance of the CNNs in pattern recognition tasks.

As shown in Fig. 10.16, LeNet has an input layer, 6 hidden layers, and an output layer. Among the 4 hidden layers, there are 2 convolutional layers, 2 subsampling (or pooling) layers, and 2 fully connected layers. Actually, common functional layers of a CNN can be e.g., fully-connected layers, convolutional layers, pooling layers, etc. It is very often that a functional layer is followed by an activation layer, such as ReLU layer, Sigmoid layer, Tanh layer, etc. After a sequence of functional and activation layers, we need a softmax layer to convert the output into a probability distribution.

In the following, we first introduce functional layers, activation functions, and softmax layer that have been widely used in various CNNs, and then present a few common practices that have been used to either prepare data for training in or support the training.

10.4.1   Functional Layers

As suggested earlier, each layer function f_i is a mapping from a high-dimensional space $\mathbb{R}^{k_{i-1}}$ (that associates with Layer-$(i-1)$) to another $\mathbb{R}^{k_i}$ (that associates with Layer-i). That is, given $\mathbf{v}_{i-1} \in \mathbb{R}^{k_i}$, we have $\mathbf{v}_i = f_i(\mathbf{v}_{i-1}) \in \mathbb{R}^{k_i}$.

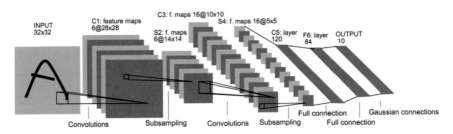

Fig. 10.16 Architecture of LeNet-5 [94], a convolutional neural network for digits recognition

Actually, in most CNN layers, the transformation f_i is conducted in two steps. For the first step, it is transformed with a linear transformation, and in the second step, every neuron passes through an activation function. Formally,

$$\mathbf{u}_i = \mathbf{W}_i \mathbf{v}_{i-1} + \mathbf{b}_i \quad \text{and} \quad \mathbf{v}_i = \sigma_i(\mathbf{u}_i) \tag{10.30}$$

where σ_i is an activation function. In the following, we introduce functional layers, followed by activation functions.

10.4.1.1 Fully-Connected Layer

In a fully connected layer, every neuron receives inputs from all neurons of the previous layer, and the output of the neuron is the result of the linear combination of the inputs. As shown in Fig. 10.17, Layer 2 is a fully-connected layer. The neuron n_{21} receives inputs from all neurons $n_{11}, \ldots, n_{16}$ in Layer 1, and weighted them with the learnable weights $\mathbf{W}_{21} = (w_{21,11}, \ldots, w_{21,16})$. Therefore, its value

$$u_{21} = \mathbf{W}_{21} \times \mathbf{v}_1 + b_2 = w_{21,11}v_{11} + \ldots + w_{21,16}v_{16} + b_2 \tag{10.31}$$

where b_2 is the bias of layer 2. Similar for the neuron n_{22}.

Note that, in the above, we use symbol u, instead of v, to denote the value of the neuron, because the output of this neuron normally needs to pass through an activation function, i.e.,

$$v_{21} = \alpha(u_{21}) \tag{10.32}$$

We will introduce activation functions α later.

In a CNN, fully connected layers often appear in the last few layers, after the convolutional layers. The main functionality of those fully-connected layers is to implement classification over those features extracted by the convolutional layers.

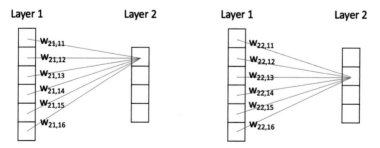

Fig. 10.17 Illustration of fully-connected layer

10.4.1.2   Convolutional Layer

Given an $n \times n$ matrix $\mathbf{x}$ and an $m \times m$ filter $\mathbf{f}$, we can compute the resulting matrix $\mathbf{z}$ by repeatedly (1) overlapping the filter over the matrix, as illustrated in Fig. 10.18 with the red dashed lines, and (2) computing an element $z_{i,j}$ with element-wise multiplication, as illustrated in Fig. 10.18 with the blue dashed lines.

The overlapping usually starts from the element $(1,1)$ of the matrix $\mathbf{x}$. Therefore, the $(1,1)$ element in the resulting matrix $\mathbf{z}$ is computed as follows with the element-wise multiplication:

$$z_{1,1} = \sum_{k=0}^{m-1} \sum_{l=0}^{m-1} x_{1+k,1+l} \times f_{1+k,1+l} \tag{10.33}$$

Afterwards, it depends on a parameter $stride$ to determine the next element on $\mathbf{x}$. For example, if $stride = 1$, then one of the next elements, along the horizontal direction, is $(1, 1 + stride) = (1, 2)$ such that

$$z_{1,2} = \sum_{k=0}^{m-1} \sum_{l=0}^{m-1} x_{1+k,1+l+stride} \times f_{1+k,1+l} = \sum_{k=0}^{m-1} \sum_{l=0}^{m-1} x_{1+k,1+l+1} \times f_{1+k,1+l} \tag{10.34}$$

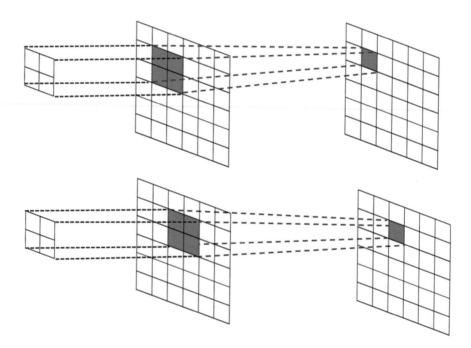

Fig. 10.18 Illustration of convolutional layer

The other next element, along the vertical direction, is $(1 + stride, 1) = (2, 1)$ such that

$$z_{2,1} = \sum_{k=0}^{m-1}\sum_{l=0}^{m-1} x_{1+k+stride,1+l} \times f_{1+k,1+l} = \sum_{k=0}^{m-1}\sum_{l=0}^{m-1} x_{1+k+1,1+l} \times f_{1+k,1+l} \tag{10.35}$$

Figure 10.18 presents the case where we move horizontally with $stride = 1$.

If $stride = 2$, the next horizontal element on $\mathbf{x}$ will be $(1, 1 + stride) = (1, 3)$ and we are computing the $(1, 2)$ element for the resulting matrix $\mathbf{z}$, i.e.,

$$z_{1,2} = \sum_{k=0}^{m-1}\sum_{l=0}^{m-1} x_{1+k,1+l+stride} \times f_{1+k,1+l} = \sum_{k=0}^{m-1}\sum_{l=0}^{m-1} x_{1+k,1+l+2} \times f_{1+k,1+l} \tag{10.36}$$

Similarly, the next vertical element $(2, 1)$ is computed as follows:

$$z_{2,1} = \sum_{k=0}^{m-1}\sum_{l=0}^{m-1} x_{1+k+stride,1+l} \times f_{1+k,1+l} = \sum_{k=0}^{m-1}\sum_{l=0}^{m-1} x_{1+k+2,1+l} \times f_{1+k,1+l} \tag{10.37}$$

Note that, no matter what the $stride$ is, the incremental to the element on $\mathbf{z}$ is always 1, to make sure that we are constructing $\mathbf{z}$ one element by one element.

We can see that, $\mathbf{z}$ is a $t \times t$ matrix such that

$$t = \frac{n - m}{stride} + 1 \tag{10.38}$$

For example, if $n = 4$ and $m = 2$, then $t = 3$ when $stride = 1$ and $t = 2$ when $stride = 2$.

10.4.1.3 Zero-Padding

As we can see from the previous discussion on the convolutional layer, the shapes of the matrices $\mathbf{x}$ and $\mathbf{z}$ are not the same. It is possible that we might be interested in maintaining the shape of the matrix along a sequence of convolutional operations. In this case, it is useful to consider a pre-processing on $\mathbf{x}$ before the convolutional filter is applied. Zero-padding, a typical pre-processing operation, is to use 0 to pad the input with 0-cells, as shown in Fig. 10.19.

4	0	1	7
5	6	9	-5
-3	8	3	6
2	-2	-1	4

0	0	0	0	0	0
0	4	0	1	7	0
0	5	6	9	-5	0
0	-3	8	3	6	0
0	2	-2	-1	4	0
0	0	0	0	0	0

Fig. 10.19 Zero-padding: pad the input with 0-cells around it

We can see that, if we pad **x** with a border of u zero valued pixels, **z** is a $t \times t$ matrix such that

$$t = \frac{n - m + 2 * u}{stride} + 1 \qquad (10.39)$$

In Fig. 10.19, $u = 1$. Therefore, if $n = 4$ and $m = 2$ and $u = 1$, then $t = 5$ when $stride = 1$ and $t = 3$ when $stride = 2$.

10.4.1.4   Pooling Layer

A pooling layer is to reduce the information in a matrix by collapsing elements with operations. The pooling layer is frequently used in convolutional neural networks with the purpose of progressively reducing the spatial size of the representation to reduce the number of features and the computational complexity of the network. Assume that, as shown in Fig. 10.20, we have a 4×4 matrix. An application of a 2×2 max-pooling filter, under the condition that $stride = 2$, will get a 2×2 matrix by collapsing every 2×2 block with the max operation.

In addition to max-pooling, there are other pooling layers, such as the average-pooling layer, which replaces the max operation with the average operation.

10.4.2   Activation Functions

As mentioned earlier, for most functional layers, they are followed by an activation layer where

10.4.2.1   ReLU

$$ReLU(x) = max(0, x) \qquad (10.40)$$

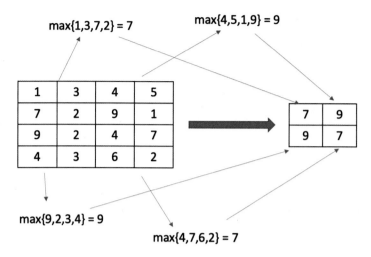

Fig. 10.20 Max-pooling. An application of 2×2 filter, $stride = 2$, on a 4×4 matrix

10.4.2.2 Sigmoid

$$\sigma(x) = \frac{1}{1 + e^{-x}} \tag{10.41}$$

such that $\sigma'(x) = \sigma(x)(1 - \sigma(x))$.

10.4.2.3 Softmax

$$\delta(\mathbf{v}) = \left(\frac{e^{v_1}}{\sum_{i=1}^{|\mathbf{v}|} e^{v_i}}, \ldots, \frac{e^{v_{|\mathbf{v}|}}}{\sum_{i=1}^{|\mathbf{v}|} e^{v_i}} \right) \tag{10.42}$$

10.4.3 Data Preprocessing

A suitable pre-processing of the training data can have a significant impact on the performance of the resulting model. In the following, we introduce a few different data pre-processing methods. Whether or not a specific pre-processing method should be applied is problem-specific, depending on the dataset and the machine learning task.

10.4.3.1   Mean Normalization

removes the mean from each data sample, i.e.,

$$\mathbf{x}' = \mathbf{x} - \bar{\mathbf{x}} \tag{10.43}$$

where $\bar{\mathbf{x}} = \dfrac{1}{|D|} \sum_{\mathbf{x} \in D} \mathbf{x}$ is the mean of the dataset D.

10.4.3.2   Standardization or Normalization

requires, on top of the mean normalisation, all features to be on the same scale, i.e., every sample $\mathbf{x}$ is converted into

$$\mathbf{x}' = \frac{\mathbf{x} - \bar{\mathbf{x}}}{\sigma_D} \tag{10.44}$$

where σ_D is the standard deviation of the dataset D.

10.4.3.3   Whitening

requires that the covariance matrix of the converted dataset is the identity matrix—1 in the diagonal and 0 for the other cells. It first applies the mean normalisation on the dataset D to get D', and then apply a whitening matrix W on every sample, i.e., let $\mathbf{x}'' = \mathbf{W}\mathbf{x}'$, such that $\mathbf{W}\mathbf{W}^T = \Sigma^{-1}$ and Σ is the non-singular covariance matrix of D'.

Depending on what $\mathbf{W}$ is, we have Mahalanobis or ZCA whitening ($\mathbf{W} = \Sigma^{-1/2}$), Cholesky whitening ($\mathbf{W} = \mathbf{L}^T$ for $\mathbf{L}$ the Cholesky decomposition of Σ^{-1}), or PCA whitening ($\mathbf{W}$ is the eigen-system of Σ^{-1}).

10.4.4   Practice

First, we setup hyper-parameters (e.g., batchsize, epoch, learning rate), device (e.g., CPU or GPU), and load training dataset (MNIST).

```
1  import torch
2  import torch.nn as nn
3  import torch.nn.functional as F
4  import torch.optim as optim
5  from torchvision import datasets, transforms
6  import argparse
7  import time
8  import os
```

```
9
10 # Setup training parameters
11 parser = argparse.ArgumentParser(description='PyTorch MNIST
      Training')
12 parser.add_argument('--batch-size', type=int, default=128,
      metavar='N',
13                       help='input batch size for training (default:
       128)')
14 parser.add_argument('--test-batch-size', type=int, default=128,
      metavar='N',
15                       help='input batch size for testing (default:
       128)')
16 parser.add_argument('--epochs', type=int, default=5, metavar='N',
17                       help='number of epochs to train')
18 parser.add_argument('--lr', type=float, default=0.01, metavar='LR
      ',
19                       help='learning rate')
20 parser.add_argument('--no-cuda', action='store_true', default=
      False,
21                       help='disables CUDA training')
22 parser.add_argument('--seed', type=int, default=1, metavar='S',
23                       help='random seed (default: 1)')
24 parser.add_argument('--model-dir', default='./model-mnist-cnn',
25                       help='directory of model for saving
      checkpoint')
26 parser.add_argument('--load-model', action='store_true', default=
      False,
27                       help='load model or not')
28
29 args = parser.parse_args(args=[])
30
31 if not os.path.exists(args.model_dir):
32     os.makedirs(args.model_dir)
33
34 # Judge cuda is available or not
35 use_cuda = not args.no_cuda and torch.cuda.is_available()
36 #device = torch.device("cuda" if use_cuda else "cpu")
37 device = torch.device("cpu")
38
39 torch.manual_seed(args.seed)
40 kwargs = {'num_workers': 1, 'pin_memory': True} if use_cuda else
      {}
41
42 # Setup data loader
43 transform=transforms.Compose([
44         transforms.ToTensor(),
45         transforms.Normalize((0.1307,), (0.3081,))
46         ])
47 trainset = datasets.MNIST('../data', train=True, download=True,
48                   transform=transform)
49 testset = datasets.MNIST('../data', train=False,
50                   transform=transform)
51 train_loader = torch.utils.data.DataLoader(trainset,batch_size=
      args.batch_size, shuffle=True,**kwargs)
```

```
52 test_loader = torch.utils.data.DataLoader(testset,batch_size=args
      .test_batch_size, shuffle=False, **kwargs)
```

We can define a convolutional neural network as follows, with 2 convolutional layers and 2 fully connected layers.

```
1  # Define CNN
2  class Net(nn.Module):
3      def __init__(self):
4          super(Net, self).__init__()
5          # in_channels:1  out_channels:32  kernel_size:3  stride:1
6          self.conv1 = nn.Conv2d(1, 32, 3, 1)
7          # in_channels:32  out_channels:64  kernel_size:3  stride
   :1
8          self.conv2 = nn.Conv2d(32, 64, 3, 1)
9          self.fc1 = nn.Linear(9216, 128)
10         self.fc2 = nn.Linear(128, 10)
11
12     def forward(self, x):
13         x = self.conv1(x)
14         x = F.relu(x)
15         x = self.conv2(x)
16         x = F.relu(x)
17         x = F.max_pool2d(x, 2)
18         x = torch.flatten(x, 1)
19         x = self.fc1(x)
20         x = F.relu(x)
21         x = self.fc2(x)
22         output = F.log_softmax(x, dim=1)
23         return output
```

```
1  # Train function
2  def train(args, model, device, train_loader, optimizer, epoch):
3      model.train()
4      for batch_idx, (data, target) in enumerate(train_loader):
5          data, target = data.to(device), target.to(device)
6
7          #clear gradients
8          optimizer.zero_grad()
9
10         #compute loss
11         loss = F.cross_entropy(model(data), target)
12
13         #get gradients and update
14         loss.backward()
15         optimizer.step()
16
17 # Predict function
18 def eval_test(model, device, test_loader):
19     model.eval()
20     test_loss = 0
21     correct = 0
22     with torch.no_grad():
23         for data, target in test_loader:
```

```
24      data, target = data.to(device), target.to(device)
25      output = model(data)
26      test_loss += F.cross_entropy(output, target,
     size_average=False).item()
27          pred = output.max(1, keepdim=True)[1]
28          correct += pred.eq(target.view_as(pred)).sum().item()
29  test_loss /= len(test_loader.dataset)
30  test_accuracy = correct / len(test_loader.dataset)
31  return test_loss, test_accuracy
```

Finally, we define the main function, which can load the trained model, or train the initial model and save the trained model.

```
1  # Main function, train the initial model or load the model
2  def main():
3      model = Net().to(device)
4      optimizer = optim.SGD(model.parameters(), lr=args.lr)
5
6      if args.load_model:
7          # Load model
8          model.load_state_dict(torch.load(os.path.join(args.
     model_dir, 'final_model.pt')))
9          trnloss, trnacc = eval_test(model, device, train_loader)
10         tstloss, tstacc = eval_test(model, device, test_loader)
11         print('trn_loss: {:.4f}, trn_acc: {:.2f}%'.format(trnloss
     , 100. * trnacc), end=', ')
12         print('test_loss: {:.4f}, test_acc: {:.2f}%'.format(
     tstloss, 100. * tstacc))
13
14     else:
15         # Train initial model
16         for epoch in range(1, args.epochs + 1):
17             start_time = time.time()
18
19             #training
20             train(args, model, device, train_loader, optimizer,
     epoch)
21
22             #get trnloss and testloss
23             trnloss, trnacc = eval_test(model, device,
     train_loader)
24             tstloss, tstacc = eval_test(model, device,
     test_loader)
25
26             #print trnloss and testloss
27             print('Epoch '+str(epoch)+': '+str(int(time.time()-
     start_time))+'s', end=', ')
28             print('trn_loss: {:.4f}, trn_acc: {:.2f}%'.format(
     trnloss, 100. * trnacc), end=', ')
29             print('test_loss: {:.4f}, test_acc: {:.2f}%'.format(
     tstloss, 100. * tstacc))
30
31             #save model
```

```
32        torch.save(model.state_dict(), os.path.join(args.
      model_dir, 'final_model.pt'))
33
34  if __name__ == '__main__':
35      main()
```

10.5   Regularisation Techniques

This section introduces several regularisation techniques, which aim to introduce inductive bias to the learning process.

Assume that, we have a model $f_{\mathbf{w}}$, being it a model whose parameters are just initialised or a model that appears during the training process. The dataset is $D = \{(\mathbf{x}_i, y_i) \mid i \in \{1..n\}\}$ is a labelled dataset. We are considering the classification task.

As we have seen in the previous chapters that most machine learning algorithms are to optimise the loss between ground truths and predictions. For example, as suggested in Eq. (7.3), the linear regression is to minimise

$$\hat{L}(f_{\mathbf{w}}) = \frac{1}{m} \sum_{i=1}^{m} (\mathbf{w}^T \mathbf{x}^{(i)} - y^{(i)})^2 \tag{10.45}$$

and, as suggested in Eq. (10.8), the convolutional neural network is to minimise

$$\hat{L}(f_{\mathbf{w}}) = \frac{1}{m} \sum_{i=1}^{m} (f_{\mathbf{w}}(\mathbf{x}^{(i)}) - y^{(i)})^2 \tag{10.46}$$

when taking the mean square error as the loss function. Based on such optimisation objectives, stochastic gradient descent based methods are applied to search for the optimal solutions. When the problem is relatively simple, e.g., the number of parameters is small, this may lead to optimal solution. However, this might not work well and it is very easy to over-fit the model when the problem is complex.

For the complex cases, it is needed to reduce the model complexity by applying regularisation techniques. In the following, we introduce a few regularisation techniques that have been widely used.

10.5.1   Ridge Regularisation

For ridge regularisation, the loss function is updated by having a penalty term, i.e.,

$$\hat{L}(f_{\mathbf{w}}) = \frac{1}{m} \sum_{i=1}^{m} (f_{\mathbf{w}}(\mathbf{x}^{(i)}) - y^{(i)})^2 + \lambda \sum_{w \in \mathbf{W}} w^2 \tag{10.47}$$

where the term $\sum_{w \in \mathbf{W}} w^2$ is the square of the magnitude of the coefficients, and λ is a hyper-parameters balancing between learning loss and the penalty term. According to the definition, the ridge regularisation reduces the model complexity and multicollinearity.

10.5.2 Lasso Regularisation

For lasso (least absolute shrinkage and selection operator) regularisation, the loss function is updated by having a penalty term, i.e.,

$$\hat{L}(f_\mathbf{w}) = \frac{1}{m} \sum_{i=1}^{m} (f_\mathbf{w}(\mathbf{x}^{(i)}) - y^{(i)})^2 + \lambda \sum_{w \in \mathbf{W}} |w| \qquad (10.48)$$

that is, instead of taking squared coefficients, we consider the absolute value of the coefficients. According to the definition, the lasso regularisation encourages the selectivity of features, i.e., make the weight matrix sparser.

10.5.3 Dropout

Dropout [154] is a regularisation technique to reduce the complex co-adaptations of training data. Essentially, it randomly ignores, or drops out, a certain percentage of the layer output during the training.

Dropout can be used on most types of layers, such as fully connected layers, convolutional layers, and the long short-term memory network layers. It may be applied to any or all hidden layers as well as the input layer, but not on the output layer. Dropout is a training technique, and is not used when making a prediction, i.e., after training.

10.5.4 Early Stopping

Early stopping is to use a holdout validation dataset to evaluate whether the training procedure should be terminated to prevent the increase of generalisation error. In general, it is applied when the performance of the model on the validation dataset starts to degrade (e.g. loss begins to increase or accuracy begins to decrease).

10.5.5 Batch-Normalisation

Batch-Normalisation is a normalisation step that fixes the means and variances of each layer's inputs. It has been shown useful for efficient training of some large scale networks.

10.6 Uncertainty Estimation

A neural network f can be seen as a probabilistic classifier because, for a classification task, given an input $\mathbf{x}$, it outputs a probabilistic distribution $f(\mathbf{x})$. That is, aleatoric uncertainty is considered. However, as suggested in Sect. 3.2, due to the existence of other uncertainties, a single probabilistic distribution is insufficient, in particular, we are not sure whether the hypothesis class (i.e., model construction) is correct and whether the training achieves the global optimal.

10.6.1 Estimating Total Uncertainty

For the epistemic uncertainty, we focus on approximation uncertainty, and assume that the model uncertainty is reduced because over-parameterised neural networks have the capacity to model any complex function. The approximation uncertainty is mainly from the weight $\mathbf{W}$. To capture this uncertainty, Bayesian neural network has been proposed. In a Bayesian neural network, every weight is represented by a probability distribution rather than a real number. Therefore, the learning of a Bayesian neural network is to compute a posterior distribution $P(\mathbf{W} \mid D_{train})$, and the predictive probability of input $\mathbf{x}$ is to compute

$$P(y \mid \mathbf{x}, D_{train}) \triangleq \int P(y \mid \mathbf{x}, \mathbf{W}) P(\mathbf{W} \mid D_{train}) d\mathbf{W} \qquad (10.49)$$

While the posterior distribution $P(\mathbf{W} \mid D_{train})$ cannot be obtained in an analytical way, it can be estimated with variational approaches, by e.g., assuming a variational distribution q with parameter θ and then minimising the KL divergence between q and $P(\mathbf{W} \mid D_{train})$ as we will discuss in Sect. 10.6.3.2. We will discuss in Sect. 10.6.3 a set of methods on how to estimate posterior distribution. We remark that, with the method suggested in Eq. (10.49), the obtained uncertainty of the predictive probability, i.e., the uncertainty of $P(y \mid \mathbf{x}, D_{train})$, is the total uncertainty, including both aleatoric and epistemic uncertainties.

For the classification task, $P(y|\mathbf{x}, \mathbf{W})$ can be the Softmax probability (or Softmax probability calibrated with techniques such as temperature scaling). Based on this and q, $P(y|\mathbf{x}, \mathbf{W})P(\mathbf{W}|D_{train})$ can be seen as re-weighting p_θ with Softmax probability. Once having the distribution $P(y|\mathbf{x}, D_{train})$, its total uncertainty can be quantified with common metrics such as entropy. For the regression task, $P(y|\mathbf{x}, \mathbf{W})$ can be predicted with the method to be discussed in Sect. 10.6.4, and in the end, the total uncertainty can be quantified with the variance of $P(y|\mathbf{x}, D_{train})$.

10.6.2   Separating Aleatoric and Epistemic Uncertainties

It is also possible to separate aleatoric and epistemic uncertainties, by taking an information-theoretical view. For example, we may use the entropy of the predictive posterior distribution

$$H[P(y \mid \mathbf{x})] \triangleq - \sum_{y \in C} P(y|\mathbf{x}) \log_2 P(y|\mathbf{x}) \tag{10.50}$$

to express the total uncertainty, and

$$\begin{aligned}
&\mathbf{E}_{P(\mathbf{W}|D_{train})} H[P(y|\mathbf{W}, \mathbf{x})] \\
&= - \int P(\mathbf{W}|D_{train}) (\sum_{y \in C} P(y|\mathbf{W}, \mathbf{x}) \log_2 P(y|\mathbf{W}, \mathbf{x})) d\mathbf{W}
\end{aligned} \tag{10.51}$$

to express the aleatoric uncertainty. Based on them, the epistemic uncertainty is the difference between them, i.e.,

$$H[P(y \mid \mathbf{x})] - \mathbf{E}_{P(\mathbf{W}|D_{train})} H[P(y|\mathbf{W}, \mathbf{x})] \tag{10.52}$$

which essentially is the mutual information between y and $\mathbf{W}$ when observing the input instance $\mathbf{x}$. Intuitively, Eq. (10.51) utilises the observation that, once fixing the weight and considering $P(y \mid \mathbf{W}, \mathbf{x})$, the epistemic uncertainty is removed from $P(y \mid \mathbf{x})$.

In practice, for the computation of aleatoric uncertainty, we need to estimate the posterior distribution $P(\mathbf{W}|D_{train})$, and for the computation of epistemic uncertainty, we need to estimate the mutual information between y and $\mathbf{W}$ for instance $\mathbf{x}$.

10.6.3   Estimating Posterior Distribution

The above estimations rely on the estimation of posterior distribution $P(W \mid D_{train})$. There are two popular methods to estimate the posterior distribution, one is the sampling method, and the other is the Laplace approximation [14, 139] of neural networks. In the following, we will sketch these two methods.

10.6.3.1   Monte Carlo Sampling

The sharpness-like method [75, 83] can be used to get a set of weight samples drawn from $(\mathbf{W} + U)$ such that $|\mathcal{L}(f_{\mathbf{W}+\mathbf{U}}) - \mathcal{L}(f_{\mathbf{W}})| \leq \epsilon$, where $U \sim \mathcal{N}(0, \sigma_U^2 I)$ is a multivariate random variable obeys zero mean Gaussian. Then, we can estimate $P(\mathbf{W} \mid D_{train})$ through these samples.

Other more advanced sampling methods, such as Markov Chain Monte Carlo and Monte Carlo (MC) dropout, can also be considered.

10.6.3.2   Variational Inference

The variational inference is to cast the computation of the distribution $P(\mathbf{W} \mid D_{train})$ as an optimisation problem. It assumes a class of tractable distributions Q and intends to finds a $q(\mathbf{W}) \in Q$ that is closest to $P(\mathbf{W} \mid D_{train})$. Apparently, once we have the distribution q, we can use it for any computation that involves $P(\mathbf{W} \mid D_{train})$.

Let

$$
\begin{aligned}
& D_{KL}(q(\mathbf{W}) \| P(\mathbf{W} \mid D_{train})) \\
&= \int q(\mathbf{W}) \log \frac{q(\mathbf{W})}{P(\mathbf{W} \mid D_{train})} d\mathbf{W} \\
&= \mathbb{E}_{q(\mathbf{W})} (\log \frac{q(\mathbf{W})}{P(\mathbf{W} \mid D_{train})}) \\
&= \mathbb{E}_{q(\mathbf{W})} (\log \frac{q(\mathbf{W})}{P(D_{train} \mid \mathbf{W}) P(\mathbf{W})} P(D_{train})) \\
&= \mathbb{E}_{q(\mathbf{W})} (\log \frac{q(\mathbf{W})}{P(D_{train} \mid \mathbf{W}) P(\mathbf{W})}) + \log P(D_{train}) \\
&= D_{KL}(q(\mathbf{W}) \| P(\mathbf{W})) - \mathbb{E}_{q(\mathbf{W})} (\log P(D_{train} \mid \mathbf{W})) + \log P(D_{train})
\end{aligned}
\tag{10.53}
$$

To minimise this, we can minimise the negative log evidence lower bound

$$
\mathcal{L}_{VI} = D_{KL}(q(\mathbf{W}) \| P(\mathbf{W})) - \mathbb{E}_{q(\mathbf{W})} (\log P(D_{train} \mid \mathbf{W}))
\tag{10.54}
$$

where $D_{KL}(q(\mathbf{W}) \| P(\mathbf{W}))$ is the KL divergence between the variational distribution $q(\mathbf{W})$ and the known prior $P(\mathbf{W})$. The expectation value $\mathbb{E}_{q(\mathbf{W})} (\log P(D_{train} \mid \mathbf{W}))$ can be approximated with Monte Carlo integration. Therefore, the optimisation

$$
\hat{q}(\mathbf{W}) \triangleq \arg \min_{q(\mathbf{W}) \in Q} \mathcal{L}_{VI}
\tag{10.55}
$$

can be conducted by iteratively improving a candidate $q(\mathbf{W})$ until convergence.

10.6.3.3   Laplace Approximation

Laplace approximation has been used in posterior estimation in Bayesian inference [12, 139]. It aims to approximate the posterior distribution $P(W|D_{train})$ by a Gaussian distribution, based on the second-order Taylor approximation of the ln posterior around its maximum-a-posteriori (MAP) estimate. Specifically, for layer l and given weights with an MAP estimate $\mathbf{W}_l^*$ on D_{train}, we have

$$
\ln P(W_l|D_{train}) \approx \ln P\left(\mathbf{W}_l^*|D_{train}\right) - \frac{1}{2}\left(W_l - \mathbf{W}_l^*\right)^T \Sigma_l \left(W_l - \mathbf{W}_l^*\right),
\tag{10.56}
$$

where

$$
\Sigma_l = \mathbb{E}_{\mathbf{x}}\left[\frac{\partial^2 \mathcal{L}(f_{\mathbf{W}}(\mathbf{x}))}{\partial \mathbf{W}_l \partial \mathbf{W}_l}\right]^{-1}
$$

is the expectation of the Hessian matrix over input data sample $\mathbf{x}$.

It is worth noting that the gradient is zero around the MAP estimate W^*, so the first-order Taylor polynomial is inexistent. Taking a closer look at Eq. (10.56), one can find that its right hand side is exactly the logarithm of the probability density function of a Gaussian distributed multivariate random variable with mean $\mathbf{W}_l^*$ and covariance Σ_l, i.e.,

$$
W_l \sim \mathcal{N}(\mathbf{W}_l^*, \Sigma_l)
\tag{10.57}
$$

where Σ_l can be viewed as the covariance matrix of W_l.

Laplace approximation suggests that it is possible to estimate Σ_l through the inverse of the Hessian matrix. Recently, [14, 139] have leveraged insights from second-order optimisation for neural networks to construct a Kronecker factored Laplace approximation. Differently from the classical second-order methods [9, 152], which suffer from high computational costs for deep neural networks, it takes advantage of the fact that Hessian matrices at the l-th layer can be Kronecker factored as explained in [107, 14]. That is,

$$
\frac{\partial^2 \mathcal{L}(f_{\mathbf{W}}(\mathbf{x}))}{\partial \mathbf{W}_l \partial \mathbf{W}_l} = \underbrace{a_{l-1} a_{l-1}^T}_{\mathcal{A}_{l-1}} \otimes \underbrace{\frac{\partial^2 \mathcal{L}(f_{\mathbf{W}}(\mathbf{x}))}{\partial h_l \partial h_l}}_{\mathcal{H}_l} = \mathcal{A}_{l-1} \otimes \mathcal{H}_l,
\tag{10.58}
$$

where h and a are the latent representation before and after the activation function, $\mathcal{A}_{l-1} \in \mathbb{R}^{N_{l-1} \times N_{l-1}}$ indicates the subspace spanned by the post-activation of the previous layer, and $\mathcal{H}_l \in \mathbb{R}^{N_l \times N_l}$ is the Hessian matrix of the loss with respect to the pre-activation of the current layer.

10.6.4 Predicting Aleatoric Uncertainty for Regression Task

Aleatoric uncertainty can be further divided into homoscedastic uncertainty and heteroscedastic uncertainty, with the former being a constant without depending on the input data and the latter depending on the input data. Because heteroscedastic uncertainty is on input data, it can be predicted as a model output. This is particularly useful for regression task, which—unlike the classification task as explained in Sect. 3.2—does not have the softmax probability that can be interpreted as the aleatoric uncertainty.

Actually, for regression task, instead of having only one output value $\hat{y}$, we may have two output values $\hat{y}$ and σ^2, where $\hat{y}$ represents the mean and σ^2 represents the variance of the output. To enable the training, the loss function can be defined as

$$\mathcal{L} = \sum_{(\mathbf{x},y) \in D_{train}} \frac{||y - \hat{y}||_2}{2\sigma^2} + \frac{1}{2} \log \sigma^2 \tag{10.59}$$

Intuitively, if the prediction is wrong, the loss function encourages to increase σ^2. On the other hand, if the prediction is close to the ground truth, the variance can be small.

10.6.5 Measuring the Quality of Uncertainty Estimation

It turns out that, while the estimation of the uncertainties is non-trivial, measuring the quality of an uncertainty estimation method or comparing the quality between two uncertainty estimation methods are harder, due to the fact that ground truth on the uncertainty is not available. However, the need for such quality measurement is compelling, because it has been known that wrongly predicted instances may be assigned with high confidence while correctly predicted ones may be assigned with low confidence. In the following, we present several methods on measuring the quality of uncertainty estimation. Unfortunately, these methods provide indicative evidence to support refuting an uncertainty estimation, without theoretical guarantees.

10.6.6 Confidence Calibration based Methods

Confidence calibration requires that the confidence score approximates the predictive probability. We use $\hat{Y}$ as a random variable to denote the predictive *label* and $\hat{P}$ as another random variable to denote the *confidence* of the prediction. A machine learning model is *perfectly calibrated* if

$$P(\hat{Y} = y \mid \hat{P} = p) = p \tag{10.60}$$

for all probability value $p \in [0, 1]$ and all label $y \in C$. Intuitively, the right-hand-side denotes a confidence value since $\hat{P} = p$, and the left-hand-side denotes the accuracy of predicting a label y given the confidence. For example, given 100 predictions, each with a confidence of 0.8, we expect that 80 should be correctly classified. The calibration error is the difference between the right-hand-side and the left-hand-side. Moreover, the model is overconfident if $P(\hat{Y} = y \mid \hat{P} = p) < p$, and under-confident if $P(\hat{Y} = y \mid \hat{P} = p) > p$.

10.6.6.1   Expected Calibration Error (ECE)

ECE discretises the probability interval into a fixed number B of bins, and assigns predictive probabilities to the bin that contains it. The calibration error becomes the difference between the fraction of predictions in the bin that are correct (i.e., accuracy) and the mean of the probabilities in the bin (i.e., confidence). Specifically, ECE computes a weighted average of this error across bins, i.e.,

$$ECE \triangleq \sum_{b=1}^{B} \frac{n_b}{N} |acc(b) - conf(b)| \tag{10.61}$$

where n_b is the number of predictions in bin b, N is the total number of data points, and $acc(b)$ and $conf(b)$ are the accuracy and confidence of bin b, respectively.

ECE is more suitable for Binary classification and can be too coarse for multi-class classification. To this end, Static Calibration Error (SCE) is proposed as a simple extension of ECE to the multiclass setting. Formally,

$$SCE \triangleq \frac{1}{|C|} \sum_{y \in C} \sum_{b=1}^{B} \frac{n_{by}}{N} |acc(b, y) - conf(b, y)| \tag{10.62}$$

where $acc(b, y)$ and $conf(b, y)$ are the accuracy and confidence of bin b for class label y, respectively, and n_{bk} is the number of predictions in bin b for class label y.

We may also consider Adaptive Calibration Error (ACE) which redefines the bin intervals so that each bin contains an equal number of predictions.

10.6.6.2   Maximum Calibration Error (MCE)

ECE considers the average quality. In high-risk applications where reliable confidence measures are necessary, we may wish to consider the worst-case deviation between confidence and accuracy. Therefore, we define

$$MCE \triangleq \max_{b \in \{1..B\}} |acc(b) - conf(b)| \tag{10.63}$$

which finds the maximal deviation among all bins.

10.6.7 Selective Prediction Method

Selective prediction requires that the confidence scores, which are obtained through uncertainty estimation methods, can be used together with the thresholds to enable the model to be abstained from making predictions on samples with low confidence scores to achieve higher accuracy on the remaining part [61]. That is, the confidence score is expected to be used for separating correct predictions and wrong predictions.

Given a threshold t and a machine learning model, we may separate a dataset D into

$$
\begin{aligned}
D_{yh} &= \{(\mathbf{x}, f(\mathbf{x})) \mid f_y(\mathbf{x}) \geq t, (\mathbf{x}, y) \in D\} \\
D_{yl} &= \{(\mathbf{x}, f(\mathbf{x})) \mid f_y(\mathbf{x}) < t, (\mathbf{x}, y) \in D\}
\end{aligned}
\tag{10.64}
$$

where we recall that $f(\mathbf{x})$ is the predictive label with f and $f_y(\mathbf{x})$ is the probability value of classifying $\mathbf{x}$ as y with the model f. Ideally, D_{yl} contains all wrong predictions and D_{yh} contains all correct predictions. Together with the dataset D which contains the ground truth labels, we can compute the TP rate, FP rate, recall, and precision, as defined in Sect. 2.4, for any given threshold t. Then, by adapting the threshold t, we may draw ROC curve and PR curve. Finally, we can measure the quality of uncertainty estimation with the metrics AUROC and AUPR, i.e., the area under ROC and PR curves.

However, it has been suggested in [31] that AUPR and AUROC may not only fail to provide a fair quality measurement, but also implicitly encourage the bad practice of reducing model accuracy in designing uncertainty estimation methods. Another curve called Risk-Coverage (RC) curve is proposed. Formally,

$$
coverage \triangleq \frac{D_h}{D}
\tag{10.65}
$$

denotes the percentage of the input processed by the model without human intervention, and

$$
riks \triangleq \mathcal{L}(f(D_h))
\tag{10.66}
$$

denotes the level of risk of these model predictions, where $\mathcal{L}$ is a loss function. Then, we can draw a curve with the *coverage* as the x-axis and the *risk* as the y-axis, and use the area under the curve, or AURC, as the measurement.

10.7 Robustness and Adversarial Attack

As explained in Sect. 3.3, an adversarial example is an input that is close enough to, but with a different predicted label with, a correctly-predicted input. In most cases, the search for an adversarial example is formalised as an optimisation problem, in a form either the same as or similar with Eq. (3.11).

10.7.1 Limited-Memory BFGS Algorithm

Some researchers [165] noticed the existence of adversarial examples, and described them as "blind spots" in DNNs. They found that adversarial images usually appear in the neighbourhood of correctly-classified examples, which can fool the DNNs although they are human-visually similar to the natural ones. It also empirically observes that random sampling in the neighbouring area (see the template solution we provided in Sect. D) is not efficient to generate such examples due to the sparsity of adversarial images in the high-dimensional space. Thus, they proposed an optimisation solution to efficiently search the adversarial examples. Formally, assume we have a classifier $f : \mathbb{R}^{s_1} \rightarrow \{1 \ldots s_K\}$ that maps inputs to one of s_K labels, and $\mathbf{x} \in \mathbb{R}^{s_1}$ is an input, $t \in \{1 \ldots s_K\}$ is a target label such that $t \neq \arg\max_l f_l(\mathbf{x})$. Then the adversarial perturbation $\mathbf{r}$ can be solved by

$$
\begin{aligned}
& \min \|\mathbf{r}\|_2 \\
s.t.\quad & \arg\max_l f_l(\mathbf{x} + \mathbf{r}) = t \\
& \mathbf{x} + \mathbf{r} \in \mathbb{R}^{s_1}
\end{aligned}
\tag{10.67}
$$

Since the exact computation is hard, an approximate algorithm based on the limited-memory Broyden–Fletcher–Goldfarb–Shanno algorithm (L-BFGS) is used instead. Furthermore, they observed that adversarial perturbations are able to transfer among different model structures and training sets, i.e., an adversarial image that aims to fool one DNN classifier also potentially deceives another neural network with different architectures or training datasets [165].

10.7.2 Fast Gradient Sign Method

Fast Gradient Sign Method [56] is able to find adversarial perturbations with a fixed L_∞-norm constraint. FGSM conducts a one-step modification to all pixel values so that the value of the loss function is increased under a certain L_∞-norm constraint. The authors claim that the linearity of the neural network classifier leads to the adversarial images because the adversarial examples are found by moving linearly along the reverse direction of the gradient of the cost function. Based on this linear explanation, an efficient linear approach is proposed to generate adversarial images [56]. Let θ represents the model parameters, $\mathbf{x}$, y denote the input and the label and $J(\theta, \mathbf{x}, y)$ is the loss function. We can calculate adversarial perturbation $\mathbf{r}$ by

$$
\mathbf{r} = \epsilon \, \text{sign} \left(\nabla_{\mathbf{x}} J(\theta, \mathbf{x}, y) \right)
\tag{10.68}
$$

A larger ϵ leads to a higher success rate of attacking, but potentially results in a bigger human visual difference. This attacking method has since been extended to a targeted and iterative version [88].

Algorithm 8: *PGDAttack*(f, $\mathbf{x}$, y, $|| \cdot ||$, d, n, ϵ), where f is the original model that the user wants to attack, $\mathbf{x}$ is the sample to be attacked, y is the true label of $\mathbf{x}$, $|| \cdot ||$ is the norm distance, d is the radius, n is the number of iterations, and $\epsilon > 0$ is an attack magnitude.

---
1 $i \leftarrow 0$
2 $\mathbf{x}^i$ is randomised such that $||\mathbf{x} - \mathbf{x}^i|| < d$
3 **repeat**
4     $\mathbf{x}^{i+1} \leftarrow Clip_{\mathbf{x}, ||\cdot||, d}(\mathbf{x}^i + \epsilon \text{sign}(\nabla_\mathbf{x} \mathcal{L}(\mathbf{x}^i, y)))$
5     $i \leftarrow i + 1$
6 **until** $i = n$;
7 **return** $\mathbf{x}^n$, as an adversarial example.

Algorithm 8 presents a pseudo code for PGD attack, where $Clip_{\mathbf{x}, ||\cdot||, d}(\mathbf{x}')$ is a clipping operation that projects any input $\mathbf{x}'$ into the norm ball centered at $\mathbf{x}$ with radius d.

10.7.3   *Jacobian Saliency Map Based Attack (JSMA)*

A L_0-norm based adversarial attacking method is also presented by exploring the *forward derivative* of a neural network [129]. Specifically, it utilises the Jacobian matrix of a DNN's logit output w.r.t. its input to identify those most sensitive pixels which then are perturbed to fool the neural network model effectively. Let c denote a target class and $\mathbf{x} \in [0, 1]^{s_1}$ represent an input image. JSMA will assign each pixel in $\mathbf{x}$ a salient weight based on the Jacobian matrix. Each salient value basically quantifies the sensitivity of the pixel to the predicted probability of class c. To generate the adversarial perturbation, the pixel with the highest salient weight is firstly perturbed by a *maximum distortion parameter* $\tau > 0$. If the perturbation leads to a mis-classification, then JSMA attack terminates. Otherwise, the algorithm will continue until a mis-classification is achieved. When a maximum L_0-norm distortion $d > 0$ is reached, the algorithm also terminates. This algorithm is primarily to produce adversarial images that are optimized under the L_0-norm distance. JMSA is generally slower than FGSM due to the computation of the Jacobian matrix.

10.7.4 DeepFool

In DeepFool, the researchers [114] introduce an iterative approach to generate adversarial images on any L_p norm distance, for $p \in [1, \infty)$. In this work, they first show how to search adversarial images for an affine binary classifier, i.e., $g(\mathbf{x}) = \text{sign}(\mathbf{w}^T \cdot \mathbf{x} + \mathbf{b})$. Given an input image $\mathbf{x}_0$, DeepFool is able to produce an optimal adversarial image by projecting $\mathbf{x}_0$ orthogonally onto the hyper-plane $\mathcal{F} = \{\mathbf{x} | \mathbf{w}^T \cdot \mathbf{x} + \mathbf{b} = 0\}$. Then this approach is generalised for a multi-class classifier: $\mathbf{W} \in \mathbb{R}^{m \times k}$ and $\mathbf{b} \in \mathbb{R}^k$. Let $\mathbf{W}_i$ and b_i be the i-th component of $\mathbf{W}$ and $\mathbf{b}$, respectively. We have

$$g(\mathbf{x}) = \underset{i \in \{1...k\}}{\text{argmax}} \, g_i(\mathbf{x}) \text{ where } g_i(\mathbf{x}) = \mathbf{W}_i^T \mathbf{x} + b_i$$

For this case, the input $\mathbf{x}_0$ is projected to the nearest face of the hyper-polyhedron P to produce the optimal adversarial image, namely,

$$P(\mathbf{x}_0) = \bigcap_{i=1}^{k} \{\mathbf{x} | g_{k_0}(\mathbf{x}) \geq g_i(\mathbf{x})\}$$

where $k_0 = g(\mathbf{x}_0)$. We can see that P is the set of the inputs with the same label as $\mathbf{x}_0$. In order to generalise DeepFool to neural networks, the authors introduce an iterative approach, namely, the adversarial perturbation is updated at each iteration by approximately linearizing the neural network and then performing the projection. Please note that, DeepFool is a heuristic algorithm for a neural network classifier that provides no guarantee to find the adversarial image with the minimum distortion, but in practice, it is a very effective attacking method.

10.7.5 Carlini & Wagner Attack

C&W Attack [19] is an optimisation based adversarial attack method which formulates finding an adversarial example as image distance minimisation problem such as L_0, L_2 and L_∞-norm. Formally, it formulates the adversarial attack as an optimisation problem to minimise

$$\mathcal{L}(\mathbf{r}) = ||\mathbf{r}||_p + c \cdot F(\mathbf{x} + \mathbf{r}), \tag{10.69}$$

where $\mathbf{x} + \mathbf{r}$ is a valid input, and F represents a surrogate function, such as $\mathbf{x} + \mathbf{r}$, which is able to fool the neural network when it is negative. The researchers directly adopt the optimiser Adam [84] to solve this optimisation problem. It is worthy to mention that C&W attack can work on three distance metrics including L_2, L_0 and L_∞ norms. A smart trick in C&W Attack lies in that it introduces a new optimisation variable to avoid box constraint (image pixel need to be within [0, 1]). C&W attack is shown to be a very strong attack which is more effective than JSMA [129], FGSM [56] and DeepFool [114]. It is able to find an adversarial example that has a significantly smaller distortion distance, especially on L_2-norm metric.

10.7.6 Adversarial Attacks by Natural Transformations

Additional to the above approaches which perform adversarial attacks at a pixel level, research has been done on crafting adversarial examples by applying natural transformations.

10.7.6.1 Rotation and Translation

Some researchers [46] argue that most existing adversarial attacking techniques generate adversarial images which appear to be human-crafted and less likely to be 'natural'. It shows that DNNs are also vulnerable to some image transformations which are likely to occur in a natural setting. For example, translating or/and rotating an input image could significantly degrade the performance of a neural network classifier. Figure 10.21 gives a few examples. Technically, given an allowed range of translation and rotation such as $\pm 3\ pixels \times \pm 30°$, the attack in [46] aims to find the minimum rotation and translation to cause a misclassification. To achieve such a purpose, in this work several ideas are explored including

- a first-order iterative method using the gradient of the DNN's loss function,
- performing an exhaustive search by discretizing the parameter space,

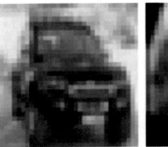

Fig. 10.21 Rotation-Translation: Original (L) 'automobile', adversarial (R) 'dog' from [46]. *The original image of an 'automobile' from the CIFAR-10 dataset is rotated (by at most 30°) and translated (by at most 3 pixels) results in an image that state-of-art classifier ResNet [60] classifies as 'dog'*

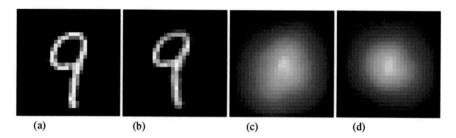

Fig. 10.22 Applying spatial transformation to MNIST image of a '9' [183]. *The Image (a) on the left is the original MNIST example image of a '9', and image (b) is the spatially transformed adversarial version that a simple convolutional network [126] labels as '8'. Notice how minor the difference between the two images is - the '9' digit has been very slightly 'bent' - but is sufficient for miss-classification. The flow-field that defines the spatial transformation is visualised in Image (c) (x-dimension) and Image (d) (y-dimension). The brighter areas indicate where the transformation is most intense - leftwards in the x-dimension and upwards in the y-dimension*

- a worst-of-k method by randomly sampling k possible parameter values and choosing the value that causes the DNN to perform the worst.

10.7.6.2 Spatially Transformed Adversarial Examples

Some researchers also introduce to produce uncontrived adversarial images via mortifying the pixel's location using spatial transformations instead of directly changing its pixel value [183]. The authors use the flow field to control the spatial transformations, which essentially quantifies the location displacement of a pixel to its new position. Figure 10.22 gives a few examples. Using a bi-linear interpolation approach the generated adversarial example is differentiable w.r.t. the flow field, which then can be formulated as an optimisation problem to calculate the adversarial flow field. Technically, they introduce a distance measure $L_{flow}(\cdot)$ (rather than the usual L_p norm distance) to capture the local geometric distortion [183]. Similar to C&W attack [19], the flow field is obtained by solving an optimisation problem in which the loss function is defined to balance between the L_{flow} loss and adversarial loss. Through human study, the attack in [183] demonstrates that adversarial examples based on such spatial transformation are more similar to original images in terms of human perceptibility, compared to those adversarial examples from L_p-norm based attacks such as FGSM [56] and C&W Attack [19].

10.7.6.3 Towards Practical Verification of Machine Learning: The Case of Computer Vision Systems (VeriVis)

The work in [132] introduces a verification framework, called VeriVis, to measure the robustness of DNNs on a set of twelve practical image transformations including reflection, translation, scale, shear, rotation, occlusion brightness, contrast, dilation, erosion, average smoothing, and median smoothing. Every transformation is controlled by a key parameter with a *polynomial-sized* domain. Those transformations are exhaustively operated on a set of input images. Then the robustness of a neural network model can be measured. VERIVIS is applied to evaluate several state-of-the-art classification models, which empirically reveals that all classifiers show a significant number of safety violations.

10.7.7   Input-Agnostic Adversarial Attacks

A key characteristic of the above attacks lies in that an adversarial example is generated with respect to a specific input, therefore cannot be applied to other inputs. Thus some researchers show more interest in *input-agnostic* adversarial perturbations.

10.7.7.1   Universal Adversarial Perturbations

The first method of input-agnostic adversarial perturbation was proposed by Moosavi-Dezfooli et al. [113], called *universal* adversarial perturbations (UAP), since UAP is able to fool a neural network on *any* input images with high probability. Let $\mathbf{r}$ the current perturbation. UAP iteratively goes through a set D of inputs sampled from the input distribution $\mathcal{D}$. At the iteration for $x_i \in D$ it updates the perturbation $\mathbf{r}$ as follows. First, it finds the minimal $\Delta \mathbf{r}_i$ w.r.t. L_2-norm distance so that $x_i + \mathbf{r} + \Delta \mathbf{r}_i$ is incorrectly classified by neural network N. Then, it projects $\mathbf{r} + \Delta \mathbf{r}_i$ back into L_p-norm ball with a radius d to enable that the generated perturbation is sufficiently small, i.e., let

$$\mathbf{r} = \arg\min_{\mathbf{r}'} ||\mathbf{r}' - (\mathbf{r} + \Delta \mathbf{r}_i)||_2$$
$$\text{s.t. } ||\mathbf{r}'||_p \leq d. \tag{10.70}$$

The algorithm will proceed until the empirical error of the sample set is sufficiently large, namely, no less than $1 - \delta$ for a pre-specified threshold δ.

10.7.7.2   Generative Adversarial Perturbations

By training a universal adversarial network (UAN), some researchers [58] generalised the C&W attack in [19] to generates input-agnostic adversarial perturbations. Assume that we have a maximum perturbation distance d and an L_p norm, a UAN $\mathcal{U}_\theta$ randomly samples an input z from a normal distribution and generates a raw perturbation $\mathbf{r}$. Then it is scaled through $\mathbf{w} \in [0, \frac{d}{||\mathbf{r}||_p}]$ to have $\mathbf{r}' = \mathbf{w} \cdot \mathbf{r}$. Then, the new input $\mathbf{x} + \mathbf{r}'$ needs to be checked with DNN N to see if it is an adversarial example. The parameters θ are optimised by adopting a gradient descent method, similar to the one used by C&W attack [19].

Later on, [137] introduced a similar approach in[58] to produce input-agnostic adversarial images. It first samples a random noise to input to UAN, and the output then is resized to meet an L_p constraint which is further added to input data, clipped, and then is used to train a classifier. This method is different to [58] in two aspects. Firstly, it explores two UAN architectures including U-Net [142] and ResNet Generator [78], and concludes ResNet Generator works better in the majority of the cases. Secondly, this work also trained a UAN by adopting several different classifiers, thus the proposed UAP can explicitly fool multiple classifiers, which is obtained by the below loss function:

$$l_{multi-fool}(\lambda) = \lambda_1 \cdot l_{fool_1} + \cdots + \lambda_m \cdot l_{fool_m} \tag{10.71}$$

where l_{fool_i} denotes the adversarial loss for classifier i, m is the number of the classifiers, and λ_i is the weight to indicate the difficulty of fooling classifier i.

10.7.8   Practice

First, we setup hyper-parameters (e.g., epsilon, steps, step size), device (e.g., CPU or GPU), and load training dataset (MNIST). Note that for FGSM: num-steps = 1 and step-size = 0.031; for PGD-20: num-steps = 20 and step-size = 0.003.

```
import torch
import torch.nn as nn
import torch.nn.functional as F
import torch.optim as optim
from torchvision import datasets, transforms
import argparse
import time
import os
from torch.autograd import Variable


# Setup training parameters
parser = argparse.ArgumentParser(description='PyTorch MNIST
    Training')
parser.add_argument('--batch-size', type=int, default=128,
    metavar='N',
                    help='input batch size for training (default:
    128)')
parser.add_argument('--test-batch-size', type=int, default=128,
    metavar='N',
                    help='input batch size for testing (default:
    128)')
parser.add_argument('--lr', type=float, default=0.01, metavar='LR
    ',
```

```
18                              help='learning rate')
19 parser.add_argument('--no-cuda', action='store_true', default=
      False,
20                              help='disables CUDA training')
21 parser.add_argument('--seed', type=int, default=1, metavar='S',
22                              help='random seed (default: 1)')
23 parser.add_argument('--model-dir', default='./model-mnist-cnn',
24                              help='directory of model for saving
      checkpoint')
25 parser.add_argument('--random', default=True,
26                              help='random initialization for PGD')
27
28
29 # FGSM: num-steps:1 step-size:0.031   PGD-20: num-steps:20 step-
      size:0.003
30 parser.add_argument('--epsilon', default=0.031,
31                              help='perturbation')
32 parser.add_argument('--num-steps', default=1,
33                              help='perturb number of steps, FGSM: 1, PGD
      -20: 20')
34 parser.add_argument('--step-size', default=0.031,
35                              help='perturb step size, FGSM: 0.031, PGD-20:
      0.003')
36
37 args = parser.parse_args(args=[])
38
39 if not os.path.exists(args.model_dir):
40     os.makedirs(args.model_dir)
41
42 # Judge cuda is available or not
43 use_cuda = not args.no_cuda and torch.cuda.is_available()
44 #device = torch.device("cuda" if use_cuda else "cpu")
45 device = torch.device("cpu")
46
47 torch.manual_seed(args.seed)
48 kwargs = {'num_workers': 1, 'pin_memory': True} if use_cuda else
      {}
49
50 # Setup data loader
51 transform=transforms.Compose([
52         transforms.ToTensor(),
53         transforms.Normalize((0.1307,), (0.3081,))
54         ])
55 trainset = datasets.MNIST('../data', train=True, download=True,
56                     transform=transform)
57 testset = datasets.MNIST('../data', train=False,
58                     transform=transform)
59 train_loader = torch.utils.data.DataLoader(trainset,batch_size=
      args.batch_size, shuffle=True,**kwargs)
60 test_loader = torch.utils.data.DataLoader(testset,batch_size=args
      .test_batch_size, shuffle=False, **kwargs)

1 # Define CNN
2 class Net(nn.Module):
```

```
 3    def __init__(self):
 4        super(Net, self).__init__()
 5        # in_channels:1  out_channels:32  kernel_size:3  stride:1
 6        self.conv1 = nn.Conv2d(1, 32, 3, 1)
 7        # in_channels:32  out_channels:64  kernel_size:3  stride
      :1
 8        self.conv2 = nn.Conv2d(32, 64, 3, 1)
 9        self.fc1 = nn.Linear(9216, 128)
10        self.fc2 = nn.Linear(128, 10)
11
12    def forward(self, x):
13        x = self.conv1(x)
14        x = F.relu(x)
15        x = self.conv2(x)
16        x = F.relu(x)
17        x = F.max_pool2d(x, 2)
18        x = torch.flatten(x, 1)
19        x = self.fc1(x)
20        x = F.relu(x)
21        x = self.fc2(x)
22        output = F.log_softmax(x, dim=1)
23        return output
```

Then we define the function for FGSM/PGD attack.

```
 1  def _pgd_whitebox(model,
 2                    X,
 3                    y,
 4                    epsilon=args.epsilon,
 5                    num_steps=args.num_steps,
 6                    step_size=args.step_size):
 7      out = model(X)
 8      err = (out.data.max(1)[1] != y.data).float().sum()
 9      X_pgd = Variable(X.data, requires_grad=True)
10      if args.random:
11          random_noise = torch.FloatTensor(*X_pgd.shape).uniform_(-
         epsilon, epsilon).to(device)
12          X_pgd = Variable(X_pgd.data + random_noise, requires_grad
         =True)
13
14      for _ in range(num_steps):
15          opt = optim.SGD([X_pgd], lr=1e-3)
16          opt.zero_grad()
17
18          with torch.enable_grad():
19              loss = nn.CrossEntropyLoss()(model(X_pgd), y)
20          loss.backward()
21          eta = step_size * X_pgd.grad.data.sign()
22          X_pgd = Variable(X_pgd.data + eta, requires_grad=True)
23          eta = torch.clamp(X_pgd.data - X.data, -epsilon, epsilon)
24          X_pgd = Variable(X.data + eta, requires_grad=True)
25          X_pgd = Variable(torch.clamp(X_pgd, 0, 1.0),
         requires_grad=True)
26      err_pgd = (model(X_pgd).data.max(1)[1] != y.data).float().sum
         ()
```

```
27    return err, err_pgd
28
29 def eval_adv_test_whitebox(model, device, test_loader):
30    # Ealuate model by white-box attack
31    model.eval()
32    robust_err_total = 0
33    natural_err_total = 0
34
35    for data, target in test_loader:
36        data, target = data.to(device), target.to(device)
37        # fgsm/pgd attack
38        X, y = Variable(data, requires_grad=True), Variable(
    target)
39        err_natural, err_robust = _pgd_whitebox(model, X, y)
40        robust_err_total += err_robust
41        natural_err_total += err_natural
42    print('natural_accuracy: {:.2f}%'.format(0.01 * (10000-
    natural_err_total)))
43    print('robust_accuracy: : {:.2f}%'.format(0.01 * (10000-
    robust_err_total)))
```

Finally, we load and attack the model.

```
1 def main():
2    model = Net().to(device)
3    model.load_state_dict(torch.load(os.path.join(args.model_dir,
     'final_model.pt')))
4    eval_adv_test_whitebox(model, device, test_loader)
5 if __name__ == '__main__':
6    main()
```

10.8   Poisoning Attack

As defined in Sect. 3.4, poisoning attack is to find a set of poisoning instances to add into the training dataset, so that the resulting trained machine learning model will perform in a wrong way according to what is required by the attacker. We consider both heuristic-based approaches, which generate poisoning samples with various heuristics from a set of base samples, and an alternating optimisation approach, which not only finds poisoning samples with heuristics but also continuously refines the obtained poisoning samples.

10.8.1   Heuristic Method

For heuristic approaches, we assume there is a set $\mathbf{X}_b$ of base samples. The objective is to compute another set $\mathbf{X}_p$ of poisoning samples such that $\mathbf{X}_p$ and $\mathbf{X}_b$ are close and a target input $\mathbf{x}_{adv}$ is classified as y_{adv}. Let g_{FE} be a feature extraction function that maps every sample into a vector of features. Practically, g_{FE} can be a neural network or part of a neural network, as discussed in Sect. 10.2.4. The heuristic methods are discussed in [150, 202, 148].

10.8.1.1 Feature Collision

Feature collision is to synthesise one poisoning sample $\mathbf{x}_p^{(i)} \in \mathbf{X}_p$ from every base sample $\mathbf{x}_b^{(i)} \in \mathbf{X}_b$ by adding perturbation. Formally,

$$\mathbf{x}_p^{(i)} \triangleq \arg\min_{\mathbf{x}} ||g_{FE}(\mathbf{x}) - g_{FE}(\mathbf{x}_{adv})||_2^2 + \beta ||\mathbf{x} - \mathbf{x}_b^{(i)}||_2^2 \tag{10.72}$$

where β is a tunable hyper-parameter to balance between two objectives. Intuitively, it requires not only the similarity between $\mathbf{x}_p^{(i)}$ and $\mathbf{x}_b^{(i)}$ but also the similarity of feature representations between $\mathbf{x}_p^{(i)}$ and the target sample $\mathbf{x}_{adv}$.

10.8.1.2 Convex Polytope

Instead of requiring the alignment of every poisoning sample's feature representation with that of target sample, it is also possible to synthesise the entire set $\mathbf{X}_p$ in a single round by requiring that $g_{FE}(\mathbf{x}_{adv})$ aligns with a convex combination of $g_{FE}(\mathbf{x}_p^{(i)})$. Formally,

$$
\begin{aligned}
\mathbf{X}_p \triangleq \quad & \arg\min_{c_i, \mathbf{x}_p^{(i)}, i=1..|\mathbf{X}_p|} ||g(\mathbf{x}_{adv}) - \sum_{i=1}^{|\mathbf{X}_p|} c_j g(\mathbf{x}_p^{(i)})||_2^2 \\
s.t. \quad & \sum_{i=1}^{|\mathbf{X}_p|} c_i = 1 \\
& \forall 1 \le i \le |\mathbf{X}_p| : c_i > 0 \\
& \forall 1 \le i \le |\mathbf{X}_p| : ||\mathbf{x}_p^{(i)} - \mathbf{x}_b^{(i)}||_\infty \le \epsilon
\end{aligned}
\tag{10.73}
$$

where $\sum_{i=1}^{|\mathbf{X}_p|} c_j g(\mathbf{x}_p^{(i)})$ is a convex combination of $g_{FE}(\mathbf{x}_p^{(i)})$ such that the coefficients $\{c_j \mid j = 1..|\mathbf{X}_p|\}$ are positive and form a probability distribution (c.f. the first two constraints). Also, we have the third constraint which ensures that the poisoning samples are close to their respective base samples.

In the following, we have two heuristic approaches for backdoor attacks. Different from the poisoning attack, we do not have a target sample $\mathbf{x}_{adv}$, but have a trigger/patch which will be added to the input instances.

10.8.1.3 Hidden Trigger Backdoor

Hidden trigger backdoor is similar with Feature Collision, except that the alignment of feature representation is not between $\mathbf{x}_p^{(i)}$ and the target sample $\mathbf{x}_{adv}$, but between $\mathbf{x}_p^{(i)}$ and a patched training image from the target class y_{adv}. Formally,

$$\mathbf{x}_p^{(i)} = \arg\min_{\mathbf{x}} ||g_{\text{FE}}(\mathbf{x}) - g_{\text{FE}}(\tilde{\mathbf{x}}_{adv}^i)||_2^2 + \beta||\mathbf{x} - \mathbf{x}_b^{(i)}||_2^2 \qquad (10.74)$$

where $\tilde{\mathbf{x}}_{adv}^i$ is patched training image from the target class y_{adv}.

10.8.1.4 Clean Label Backdoor

Clean label backdoor proceeds in two steps. First, it generates an adversarial example $\tilde{\mathbf{x}}_{adv}^i$ for each base sample. Second, it constructs $\mathbf{x}_{adv}^i$ by adding a patch to $\tilde{\mathbf{x}}_{adv}^i$.

10.8.2 An Alternating Optimisation Method

As discussed in Sect. 3.4, the data poisoning attack is a bi-level optimisation problem, which cannot be solved in a tractable way when the objective functions $\mathcal{L}_{adv}$ and $\mathcal{L}_{train}$ are non-convex. Algorithm 9 presents an algorithm that solves the problem in an alternating optimisation way. Another method to solve the bi-level optimisation is discussed in [28].

It repeats by first conducting a line search over the gradient $\nabla_{\mathbf{x}_c}\mathcal{L}_{adv}(\mathbf{x}_{adv}, y_{adv}; \mathbf{W}^i)$ for all candidate samples $\mathbf{x}_c$ in $\mathbf{X}_p$ (Line 7–10), and then conducting a standard training over the updated $\mathbf{X} \cup \mathbf{X}_p$ (Line 11). The algorithm terminates when the loss over the target input converges with respect to a pre-specified threshold ϵ (Line 14). Finally, it returns the set of updated poisoning samples (Line 15).

According to the above discussion, the key difficulty is on the computation of

$$\nabla_{\mathbf{x}_c}\mathcal{L}_{adv}(\mathbf{x}_{adv}, y_{adv}; \mathbf{W}^i) \qquad (10.75)$$

which depends not only on the input $\mathbf{x}_c$ but also on the weight $\mathbf{W}^i$. Therefore, we can apply the chain rule and get

$$\nabla_{\mathbf{x}_c}\mathcal{L}_{adv} = \nabla_{\mathbf{x}_c}\mathbf{W}^T \cdot \nabla_{\mathbf{W}}\mathcal{L}_{adv} \qquad (10.76)$$

where we have the weight $\mathbf{W}$ depends on $\mathbf{x}_c$.

It is noted that $\nabla_{\mathbf{W}}\mathcal{L}_{adv}$ can be obtained through the neural network $f_{\mathbf{W}}$ over the loss $\mathcal{L}_{adv}$. In the following, we explain how to compute $\nabla_{\mathbf{x}_c}\mathbf{W}^T$. To do so, we

---

Algorithm 9: *PoisoningAttack*($\mathbf{x}_{adv}$, y_{adv}, $\mathbf{X}_p$, $\mathbf{X}$, $\mathcal{L}_{adv}$, $\mathcal{L}_{train}$, ϵ), where $\mathbf{x}_{adv}$ is the target input, y_{adv} is the target label, $\mathbf{X}_p$ is a set of initial poisoning samples, $\mathbf{X}$ is the training dataset, $\mathcal{L}_{adv}(\mathbf{x}_{adv}, y_{adv}; \mathbf{W})$ is a function to measure the accuracy of predicting $\mathbf{x}_{adv}$ as y_{adv} with a neural network whose parameters are $\mathbf{W}$, $\mathcal{L}_{train}(\mathbf{X}, \mathbf{y})$ is the standard training loss function, and $\epsilon > 0$ is a threshold that will be used to determine the convergence.

---

1 $i \leftarrow 0$
2 $\mathbf{X}_p^i \leftarrow \mathbf{X}_p$
3 $\mathbf{W}^i \leftarrow \arg\min_{\mathbf{W}} \mathcal{L}_{train}(\mathbf{X} \cup \mathbf{X}_p^i, \mathbf{y}; \mathbf{W})$, where $\mathbf{y}$ includes labels for both $\mathbf{X}$ and $\mathbf{X}_p^i$
4 $l^i \leftarrow \mathcal{L}_{adv}(\mathbf{x}_{adv}, y_{adv}; \mathbf{W}^i)$
5 **repeat**
6 $\quad$ $\mathbf{X}_p^{i+1} \leftarrow \emptyset$
7 $\quad$ **for** $c = 1 \ldots |X_p|$ **do**
8 $\quad\quad$ $\mathbf{x}_c^{i+1} \leftarrow line\_search(\mathbf{x}_c^i, \nabla_{\mathbf{x}_c} \mathcal{L}_{adv}(\mathbf{x}_{adv}, y_{adv}; \mathbf{W}^i))$
9 $\quad\quad$ $\mathbf{X}_p^{i+1} \leftarrow \mathbf{X}_p^{i+1} \cup \{\mathbf{x}_c^{i+1}\}$
10 $\quad$ **end**
11 $\quad$ $\mathbf{W}^{i+1} \leftarrow \arg\min_{\mathbf{W}} \mathcal{L}_{train}(\mathbf{X} \cup \mathbf{X}_p^{i+1}, \mathbf{y}; \mathbf{W}^i)$
12 $\quad$ $l^{i+1} \leftarrow \mathcal{L}_{adv}(\mathbf{x}_{adv}, y_{adv}; \mathbf{W}^{i+1})$
13 $\quad$ $i \leftarrow i + 1$
14 **until** $l^i - l^{i-1} < \epsilon$;
15 **return** the final $\mathbf{X}_p^i$

---

require that $\nabla_{\mathbf{W}} \mathcal{L}_{train}(\mathbf{X} \cup \mathbf{X}_p, \mathbf{y}; \mathbf{W}) = 0$, according to the Karush-Kuhn-Tucker (KKT) equilibrium conditions, and further such conditions to remain valid while updating $\mathbf{x}_c$, i.e.,

$$\nabla_{\mathbf{x}_c}(\nabla_{\mathbf{W}} \mathcal{L}_{train}(\mathbf{X} \cup \mathbf{X}_p, \mathbf{y}; \mathbf{W})) = 0 \tag{10.77}$$

Through the application of chain rule on the above equation, we have

$$\nabla_{\mathbf{x}_c} \nabla_{\mathbf{W}} \mathcal{L}_{train} + \nabla_{\mathbf{x}_c} \mathbf{W}^T \cdot \nabla_{\mathbf{W}}^2 \mathcal{L}_{train} = 0 \tag{10.78}$$

Therefore, we have

$$\nabla_{\mathbf{x}_c} \mathbf{W}^T = -\nabla_{\mathbf{x}_c} \nabla_{\mathbf{W}} \mathcal{L}_{train}(\nabla_{\mathbf{W}}^2 \mathcal{L}_{train})^{-1} \tag{10.79}$$

where $\nabla_{\mathbf{W}} \mathcal{L}_{train}$ is obtained from the neural network $f_{\mathbf{W}}$ over the loss $\mathcal{L}_{train}$.

10.9 Model Stealing

As described in Sect. 3.5, one of the typical model stealing attacks is to construct another model $f_{surrogate}$ that is *functionally equivalent* to the victim model f_{victim}.

We assume the black-box knowledge to the victim model, which is realistic for MLaaS applications. Specifically, for a given instance, only the output probability vector is available, or more restrictively, only the predictive label is available.

For an attack, most existing methods take three steps to construct $f_{surrogate}$:

1. First, we build an architecture, and train an initial model $f_{surrogate}$ from easily accessible data such as randomly sampled data or public data.
2. Second, we synthesise further instances $\mathbf{X}_{syn}$ that are on the data distribution of the victim model.
3. Third, we update the model $f_{surrogate}$ with the new training instances $\mathbf{X}_{syn}$.

Moreover, considering that a single execution of the above steps might not achieve the best result, it is often that the last two steps are repeated until a pre-specified termination condition is satisfied.

In the following, we first present a simple iterative algorithm, and then discuss alternative implementations to the individual steps.

10.9.1   An Iterative Algorithm

In this section, we suggest a simple, iterative algorithm that learns the model $f_{surrogate}$ gradually until convergence. The algorithm is presented in Algorithm 10. We take the cross-entropy loss, $\mathcal{L}_{CE}$, as the training loss, but the algorithm can be extended to work with other loss functions.

The algorithm proceeds by first randomly selecting a synthetic dataset $\mathbf{X}_{syn}$ (Line 1) to train a surrogate model $f_{surrogate}$ (Line 2). Then, these two objects, $\mathbf{X}_{syn}$ and $f_{surrogate}$, are iteratively updated until the difference between $f_{surrogate}$ and f_{victim} is smaller than the threshold ϵ (Line 3). The iterative update is conducted by synthesising samples one by one for $\mathbf{X}_{syn}$ (Line 6–11). For every sample, its synthesis process proceeds by first randomly sampling a label (represented as a vector of probability values $\mathbf{y}$) (Line 7), and then according to the label $\mathbf{y}$ finding an instance $\mathbf{x}$ with the minimum loss on the surrogate model $f_{surrogate}$ (Line 8).

10.9.2   Initiating Surrogate Model

The initial training data can be, as in Algorithm 10, a set of instances randomly sampled from the victim model. Besides, for MLaaS, it is often useful to consider public datasets such as in [122, 123]. For the initial model, it can start from a self-constructed architecture with randomised weights, such as in [128], but can also start from pre-trained models from model zoos, such as Caffe Model Zoo [73].

Algorithm 10: *ModelStealingAttack*(f_{victim}, ϵ, n), where f_{victim} is the victim model that the user can access/query, $\epsilon > 0$ is a threshold that will be used to determine the convergence, and n is the number of samples in the synthesised dataset.

---

1 randomly sample a dataset $\mathbf{X}_{syn}$

2 $f_{surrogate} \leftarrow \underset{f}{\arg\min} \sum_{\mathbf{x} \in \mathbf{X}_{syn}} \mathcal{L}_{CE}(f(\mathbf{x}), f_{victim}(\mathbf{x}))$

3 while $\dfrac{1}{|\mathbf{X}_{syn}|} \sum_{\mathbf{x} \in \mathbf{X}_{syn}} \mathcal{L}_{CE}(f_{surrogate}(\mathbf{x}), f_{victim}(\mathbf{x})) \geq \epsilon$ **do**

4 $i \leftarrow 0$

5 $\mathbf{X}_{syn} = \emptyset$

6   **repeat**

7   sample a vector $\mathbf{y}$ as the output prediction of a candidate input

8 $\mathbf{x} \leftarrow \underset{\mathbf{x}}{\arg\min} \mathcal{L}_{CE}(f_{surrogate}(\mathbf{x}), \mathbf{y})$

9 $\mathbf{X}_{syn} \leftarrow \mathbf{X}_{syn} \cup \{\mathbf{x}\}$

10 $i \leftarrow i + 1$

11   **until** $i > n$;

12 $f_{surrogate} \leftarrow \underset{f}{\arg\min} \sum_{\mathbf{x} \in \mathbf{X}_{syn}} \mathcal{L}_{CE}(f(\mathbf{x}), f_{victim}(\mathbf{x}))$

13 end

14 return $f_{surrogate}$

10.9.3   Synthesising Further Instances

Usually, the training dataset for the initial surrogate model is much smaller than the victim model's training dataset, and for this reason, the initial model is not close enough to the victim model. Therefore, it is desirable to actively synthesise more instances for the update of the model. There are different approaches on how to get the new instances, including from public dataset [122, 123], from adversarial examples [128, 189], and from data distribution of the victim model [55].

10.9.4   Updating Surrogate Model

The update can be done by minimising the difference between $f_{surrogate}$ and f_{victim} over the newly synthesised data instances, as shown in Line 12 of Algorithm 10. We can also use a similar condition as the termination condition of the iterative process, as shown in Line 3 of Algorithm 10.

10.10 Membership Inference

As described in Sect. 3.6, membership inference is to determine if an input $\mathbf{x}$ is in the training dataset D_{train} of a machine learning model f. In the following, we consider two classes of attack methods: metric-based methods and binary classifier based methods. While these methods may have different formalism and performance, all of them are based on the fact that the machine learning model f has different behaviour on training data and test data. The behavioural difference is exhibited through e.g., prediction probability, loss, and hidden activations.

10.10.1 Metric Based Method

Metric based method proceeds by first calculating a metric M and then determining the membership of $\mathbf{x}$ by comparing the value of the metric with a pre-specified threshold. According to the knowledge available to the attacker, there are different metrics as follows.

10.10.1.1 Prediction Label Metric

Recall that C is the set of classes and y is the ground-truth label of $\mathbf{x}$, we have

$$M_{label}(\mathbf{x}) = \mathbb{1}(\arg\max_{c \in C} f(\mathbf{x})(c) = y) \tag{10.80}$$

where $\hat{y} = \arg\max_{c \in C} f(\mathbf{x})(c)$ is the predictive label and $\mathbb{1}$ is the indicator function such that

$$\mathbb{1}(a) = \begin{cases} 1 \text{ if } a \text{ is True} \\ 0 \text{ otherwise} \end{cases} \tag{10.81}$$

Intuitively, this metric is based on the assumption that the model f generalises so badly that it can only predict correctly on the training dataset. Based on this assumption, the sample is in the training dataset if and only if the prediction made by f is correct. This attack method is black-box, i.e., it requires only knowledge **K6** (Sect. 3.7).

However, most machine learning algorithms—in particular deep neural networks—can generalise well, which renders the above metric too coarse for membership inference.

10.10.1.2 Prediction Loss Metric

Let $\mathcal{L}$ be the training loss function and ϵ a prespecified threshold, we define

$$M_{loss}(\mathbf{x}) = \mathbb{1}(\mathcal{L}(f(\mathbf{x}), y) \leq \epsilon) \qquad (10.82)$$

Intuitively, it means that the sample $\mathbf{x}$ is a member of the training dataset if its prediction loss is insignificant (i.e., smaller than ϵ). The rationale behind this is that, the machine learning algorithm is trained by minimising the loss of the training data, and therefore a smaller loss suggests a higher possibility that the sample is in the training dataset. Unlike M_{label} which is a black-box attack, this method requires the knowledge about the training process (or more specifically, the training loss function), i.e., the attacker knowledge **K5**.

10.10.1.3 Prediction Entropy Metric

Recall the entropy as defined in Definition 5.1, we have

$$M_{loss}(\mathbf{x}) = \mathbb{1}(H(f(\mathbf{x})) \leq \epsilon) \qquad (10.83)$$

where ϵ is a pre-specified threshold. This metric is based on an observation that the uncertainty of the prediction, expressed as the entropy of the prediction probability, is smaller for training data sample than test data sample. This method is black-box.

10.10.2 Binary Classifier Based Method

While the above metric-based methods are simple and easy to compute, they might not perform well because modern machine learning algorithms can generalise well, and more importantly, the machine learning models are very complex so that their different behaviours on the test and training datasets cannot be easily differentiated with simple metrics. This has led to the below discussion on learning a binary classifier for the membership inference.

Intuitively, Binary classifier based attack is to learn a binary classifier f_a, which is able to predict the membership according to the prediction probability of the victim model f_{victim}. Algorithm 11 presents a generic framework for conducting the membership inference attack. It is a black-box algorithm and therefore can be applied to any machine learning model.

The algorithm takes three steps to train a binary classifier f_a, as discussed in the following. The first step is to synthesise datasets $\mathbf{X}_{syn,train}$ and $\mathbf{X}_{syn,test}$ that are of the same distribution as the training and test dataset of f_{victim}, respectively. This can be obtained with e.g., the iterative process as in Line 6–11 of Algorithm 10. But other algorithms may also apply, for example. we can first learn a distribution from

---

Algorithm 11: *MembershipInferenceAttack*(f_{victim}, m, k), where f_{victim} is the original model that the user can access/query, m is the number of input features, and k is the number of classes.

---

1 Synthesise datasets $\mathbf{X}_{syn,train}$ and $\mathbf{X}_{syn,test}$ such that $\mathbf{X}_{syn,train} \cap \mathbf{X}_{syn,test} = \emptyset$ and they follow the same distributions as the training and testing datasets of f_{victim}, respectively.
2 Construct a function $f_g : \mathbb{R}^m \to (\mathbb{R}^k, \{0, 1\})$ that maps instances in $\mathbf{X}_{syn,train}$ and $\mathbf{X}_{syn,test}$ to pairs of (probability distribution, Binary value)
3 Train a Binary classifier $f_a : \mathbb{R}^k \to \{0, 1\}$ that maps probability distributions to binary values by using the dataset $f_g(\mathbf{X}_{syn,train}) \cup f_g(\mathbf{X}_{syn,test})$.
4 **return** $f_a \cdot f_{victim} : \mathbb{R}^m \to \{0, 1\}$, which maps any sample $\mathbf{x}$ to a binary value indicating whether $\mathbf{x}$ is in the training dataset of f_{victim}.

---

the training dataset with e.g., a generative model, and then sample dataset $\mathbf{X}_{syn,train}$ from the learned distribution.

The second step is to construct a function f_g over the datasets $\mathbf{X}_{syn,rain}$ and $\mathbf{X}_{syn,test}$. This is done by first using $\mathbf{X}_{syn,train}$ to train a surrogate model $f_{surrogate}$, and then letting

$$f_g(\mathbf{x}) = \begin{cases} (f_{surrogate}(\mathbf{x}), 1) & \text{if } \mathbf{x} \in \mathbf{X}_{syn,train}, \\ (f_{surrogate}(\mathbf{x}), 0) & \text{if } \mathbf{x} \in \mathbf{X}_{syn,test} \end{cases} \tag{10.84}$$

The third step is to train a Binary classifier f_a. From f_g and $\mathbf{X}_{syn} = \mathbf{X}_{syn,train} \cup \mathbf{X}_{syn,test}$, we have a dataset $f_g(\mathbf{X}_{syn})$. Then, the Binary classifier f_a can be obtained by using $f_g(\mathbf{X}_{syn})$ as the training dataset on any machine learning model.

10.10.2.1   Predicting Whether New Sample x is in the Training Dataset of f_{victim}

If $f_a(f_{victim}(\mathbf{x})) = 0$ then it is predicted that $\mathbf{x}$ is not in the training dataset. However, if $f_a(f_{victim}(\mathbf{x})) = 1$ then it is predicted that $\mathbf{x}$ is in the training dataset.

10.10.2.2   Enhancement of f_a Through Ensemble Method

The computation of f_a can be improved if we take ensemble method. That is, we can have a set of classifiers $f_{a,1}, \ldots, f_{a,k}$ with Algorithm 11 and then the prediction on an input $\mathbf{x}$ is done through a voting mechanism over these classifiers. Alternatively, we can repeat the first two steps to have a set of datasets $f_{g,1}(\mathbf{X}_{syn,1}), \ldots, f_{g,k}(\mathbf{X}_{syn,k})$, and then train a Binary classifier f_a over the joint dataset $f_{g,1}(\mathbf{X}_{syn,1}) \cup \ldots \cup f_{g,k}(\mathbf{X}_{syn,k})$.

10.10.2.3   White-Box Attack

If the attacker has more knowledge of the model f_{victim}, we can let the surrogate models have the same structure as the victim model and then lift the first element of $f_g(\mathbf{x})$ to include not only the prediction probability but also other information such as the latent representations of layers. In this way, the Binary classifier f_a will have more input features, which will lead to more accurate membership inference.

10.10.2.4   Discussion

The above Binary classifier based membership inference algorithm is generic and model agnostic. It works well for simple datasets, such as low-dimensional tabular datasets, but might not work well for complex datasets, such as high-resolution image datasets. This is mainly because of the creation of dataset $\mathbf{X}_{syn}$. For high-dimensional datasets, the creation of a synthetic dataset that is of the same distribution as the training dataset is non-trivial, and may require a large number of samples.

Moreover, a membership inference attack becomes easier when the original model is overfitted. Intuitively, an overfitted model "remembers" the training data samples in its trainable parameters, and is therefore subject to the attack. Therefore, the improvement to the generalisation—or the reduction of generalisation error— of the model also poses a positive impact on data privacy. Nevertheless, a more principled study on the exact relation between them is needed.

10.11   Model Inversion

We consider the case of reconstructing an instance $\mathbf{x}$ based on the predictive output probability $f(\hat{\mathbf{x}})$. There are mainly two classes of attacks: optimisation based attacks and training based attacks, which we will briefly discuss below.

10.11.1   Optimisation Based Method

The basic idea is to apply gradient based search in the input domain $\mathcal{D}$ to find an instance $\mathbf{x}$ whose predictive output probability is close to a pre-specified probability vector $f(\hat{\mathbf{x}})$. Formally, it is to solve the below optimisation problem:

$$\arg\min_{\mathbf{x}} \mathcal{L}(f(\mathbf{x}), f(\hat{\mathbf{x}})) \tag{10.85}$$

However, a naive optimisation process like Eq. (10.85) tends to generate instances (such as images) that do not really resemble natural input. Therefore, for image classification tasks, some image priors have been proposed as regularisers, i.e., we may consider the following optimisation problem:

$$\arg\min_{\mathbf{x}} \mathcal{L}(f(\mathbf{x}), f(\hat{\mathbf{x}})) + P(\mathbf{x}) \tag{10.86}$$

where $P(\mathbf{x})$ can be e.g., the norm distance metric such as $||\mathbf{x}||_2^2$ [153] or $||\mathbf{x}||_6^6$ [105], or total variation [105]

$$\sum_{i,j} ((x_{i+1,j} - x_{i,j})^2 + (x_{i,j+1} - x_{i,j})^2)^{\beta/2}, \tag{10.87}$$

or a combination of these priors [188].

We remark that, the optimisation based approaches require explicit access to the gradient, and therefore are white-box attacks. Moreover, they need to work with individual instances, and for this reason, can be time-consuming when there are many instances to invert.

10.11.2   Training Based Method

Instead of optimisation based approaches, we may consider training based approaches, which aim to train a model g such that

$$\arg\min_{g} \mathbb{E}_{\mathbf{x}\in\mathcal{D}}\mathcal{L}(g(f(\mathbf{x})), \mathbf{x}) \tag{10.88}$$

where $\mathcal{L}$ is a loss function to measure the similarity of two instances, such as the L_2 norm distance. Intuitively, g can invert the model f such that the results, expressed with $g(f(\mathbf{x}))$, is close to the original input instance $\mathbf{x}$.

Recall that $f : \mathbb{R}^{s_1} \to \mathbb{R}^{s_K}$. Note that, $g : \mathbb{R}^{s_K} \to \mathbb{R}^{s_1}$ is a function mapping from the output of f to the input domain. Similar to the synthesis of instances as in model stealing (Chap. 10.9), the training dataset for g can be obtained either from public datasets or random sampling of the model f. Some additional operation may be applied to the training dataset before it is used for the training of g [187]. While the training of g may take up certain resources, it can invert instances efficiently without having prior knowledge about the model f.

Exercises

Question 1 Consider the example dataset in Example 5.1, please compute the below expressions (up to 2 decimal places)

- *GainRatio*(*Humidity*, *PlayTennis*) = 0.16
- *GainRatio*(*Temperature*, *PlayTennis*) = 0.03
- *GainRatio*(*Wind*, *PlayTennis*) = 0.05
- *InfoGain*(*Humidity*, *PlayTennis*) = 0.15
- *InfoGain*(*Temperature*, *PlayTennis*) = 0.03
- *InfoGain*(*Wind*, *PlayTennis*) = 0.05   □

Question 2 Consider part of the **Iris** dataset in Table 10.1, please compute the below expressions (up to 2 decimal places)

- *GainRatio*(*SepalLength*, *IrisClass*) =
- *GainRatio*(*SepalWidth*, *IrisClass*) =
- *GainRatio*(*PedalLength*, *IrisClass*) =
- *GainRatio*(*PedalWidth*, *IrisClass*) =
- *InfoGain*(*SepalLength*, *IrisClass*) =
- *InfoGain*(*SepalWidth*, *IrisClass*) =
- *InfoGain*(*PedalLength*, *IrisClass*) =
- *InfoGain*(*PedalWidth*, *IrisClass*) =

Question 3 Understand the basic idea of random forest by conducting research on the literature, and implement a random forest algorithm based on the decision tree algorithm to see if random forest performs better than decision tree on **Iris** dataset.   □

Question 4 Following the last newquestion, please give an adversarial attack algorithm for random forest.   □

Question 5 Consider the four data samples in Example 7.1 (also provided in Table 10.2) and the mean square error, if we have the following two functions:

Table 10.1 Part of **Iris** dataset

Index	Sepal length	Sepal width	Pedal length	Pedal width	Iris class
1	5	3	1	0.5	0
2	4	2	1	0.5	0
3	4	3	1	0.5	0
4	5	3	1	0.5	0
5	4	3	1	0.5	0
6	7	3	4	1	1
7	6	3	4	1	1
8	6	3	4	1	1
9	4	2	3	1	1
10	6	3	6	2	2
11	5	2	5	2	2
12	7	3	5	2	2
13	5	2	5	2	2
14	7	2	5	1	2

Table 10.2 A small dataset

X_1	X_2	X_3	Y
182	87	11.3	325
189	92	12.3	344
178	79	10.6	350
183	90	12.7	320

- $f_{\mathbf{w}_1} = 2X_1 + 1X_2 + 20X_3 - 330$
- $f_{\mathbf{w}_2} = X_1 - 2X_2 + 23X_3 - 332$

please newanswer the following newquestions:

1. which model is better for linear regression?
2. which model is better for linear classification by considering 0-1 loss for $\mathbf{y}^T = (0, 1, 1, 0)$?
3. which model is better for logistic regression for $\mathbf{y}^T = (0, 1, 1, 0)$?
4. According to the logistic regression of the first model, what is the prediction result of the first model on a new input $(181, 92, 12.4)$? □

Answer 1. Because

$$f_{\mathbf{w}_1}(\mathbf{x}_1) - y_1 = 2 * 182 + 1 * 87 + 20 * 11.3 - 330 - 325 = 22$$
$$f_{\mathbf{w}_1}(\mathbf{x}_2) - y_2 = 2 * 189 + 1 * 92 + 20 * 12.3 - 330 - 344 = 42$$
$$f_{\mathbf{w}_1}(\mathbf{x}_3) - y_3 = 2 * 178 + 1 * 79 + 20 * 10.6 - 330 - 350 = -33$$
$$f_{\mathbf{w}_1}(\mathbf{x}_4) - y_4 = 2 * 183 + 1 * 90 + 20 * 12.7 - 330 - 320 = 60$$
$$f_{\mathbf{w}_2}(\mathbf{x}_1) - y_1 = 182 - 2 * 87 + 23 * 11.3 - 332 - 325 \quad = -389.1$$
$$f_{\mathbf{w}_2}(\mathbf{x}_2) - y_2 = 189 - 2 * 92 + 23 * 12.3 - 332 - 344 \quad = -388.1$$
$$f_{\mathbf{w}_2}(\mathbf{x}_3) - y_3 = 178 - 2 * 79 + 23 * 10.6 - 332 - 350 \quad = -418.2$$
$$f_{\mathbf{w}_2}(\mathbf{x}_4) - y_4 = 183 - 2 * 90 + 23 * 12.7 - 332 - 320 \quad = -356.9$$
$$(10.1)$$

the model $f_{\mathbf{w}_1}$ is better for linear regression, according to the loss function (Eq. (7.3));

2. Because

$$step(f_{\mathbf{w}_1}(\mathbf{x}_1)) = 1$$
$$step(f_{\mathbf{w}_1}(\mathbf{x}_2)) = 1$$
$$step(f_{\mathbf{w}_1}(\mathbf{x}_3)) = 1$$
$$step(f_{\mathbf{w}_1}(\mathbf{x}_4)) = 1$$
$$step(f_{\mathbf{w}_2}(\mathbf{x}_1)) = 0$$
$$step(f_{\mathbf{w}_2}(\mathbf{x}_2)) = 0$$
$$step(f_{\mathbf{w}_2}(\mathbf{x}_3)) = 0$$
$$step(f_{\mathbf{w}_2}(\mathbf{x}_4)) = 0$$
$$(10.2)$$

we have that both models are the same, according to the loss function (Eq. (7.13));

3. Because

$$y_1 * \log(\sigma(f_{\mathbf{w}_1}(\mathbf{x}_1))) + (1 - y_1)\log((1 - \sigma(f_{\mathbf{w}_1}(\mathbf{x}_1)))) = -M$$
$$y_2 * \log(\sigma(f_{\mathbf{w}_1}(\mathbf{x}_2))) + (1 - y_2)\log((1 - \sigma(f_{\mathbf{w}_1}(\mathbf{x}_2)))) = 0$$
$$y_3 * \log(\sigma(f_{\mathbf{w}_1}(\mathbf{x}_3))) + (1 - y_3)\log((1 - \sigma(f_{\mathbf{w}_1}(\mathbf{x}_3)))) = 0$$
$$y_4 * \log(\sigma(f_{\mathbf{w}_1}(\mathbf{x}_4))) + (1 - y_4)\log((1 - \sigma(f_{\mathbf{w}_1}(\mathbf{x}_4)))) = -M$$
$$y_1 * \log(\sigma(f_{\mathbf{w}_2}(\mathbf{x}_1))) + (1 - y_1)\log((1 - \sigma(f_{\mathbf{w}_2}(\mathbf{x}_1)))) = 0$$
$$y_2 * \log(\sigma(f_{\mathbf{w}_2}(\mathbf{x}_2))) + (1 - y_2)\log((1 - \sigma(f_{\mathbf{w}_2}(\mathbf{x}_2)))) = -44.1$$
$$y_3 * \log(\sigma(f_{\mathbf{w}_2}(\mathbf{x}_3))) + (1 - y_3)\log((1 - \sigma(f_{\mathbf{w}_2}(\mathbf{x}_3)))) = -68.2$$
$$y_4 * \log(\sigma(f_{\mathbf{w}_2}(\mathbf{x}_4))) + (1 - y_4)\log((1 - \sigma(f_{\mathbf{w}_2}(\mathbf{x}_4)))) = -1.1$$
$$(10.3)$$

where M represents a large number, so we have

$$\hat{L}(f_{\mathbf{w}_1}) = -\tfrac{1}{4}(-M + 0 + 0 - M) = M/2$$
$$\hat{L}(f_{\mathbf{w}_2}) = -\tfrac{1}{4}(0 - 44.1 - 68.2 - 1.1) = 28.35$$
$$(10.4)$$

according to Eq. (7.18). Therefore, $f_{\mathbf{w}_2}$ is better.

4. According to Eq. (7.16), we have

Table 10.3 A small dataset

Gender	HrsWorked	Wealthy?
F	39	Y
F	45	N
M	35	N
M	43	N
F	32	Y
F	47	Y
M	34	Y

$$P_{\mathbf{w}_1}(y = 1|\mathbf{x}) = \sigma(2 * 181 + 1 * 92 + 20 * 12.4 - 330) = 1$$
$$P_{\mathbf{w}_1}(y = 0|\mathbf{x}) = 1 - \sigma(2 * 181 + 1 * 92 + 20 * 12.4 - 330) = 0$$
(10.5)

Therefore, it is predicted to 1.   □

Question 6 Understand the basic idea of Bayesian linear regression by conducting research on the literature, and implement a Bayesian linear regression algorithm to compare its performance with linear regression.   □

Question 7 Write a program for the adversarial attack for logistic regression.   □

Question 8 Given a function $f(x) = e^x/(1 + e^x)$, how many critical points?   □

Answer 0   □

Question 9 Given a function $f(x_1, x_2) = 9x_1^2 + 3x_2 + 4$, how many critical points?
                                                                                           □

Answer 0, because there is no assignment to x_1 and x_2 that can make the gradient of $f(x_1, x_2)$ equal to 0.   □

Question 10 Consider the dataset in Table 10.3, please newanswer the following newquestion:

- $P(Wealthy = Y) = 4/7$
- $P(Wealthy = N) = 3/7$
- $P(Gender = F|Wealthy = Y) = 3/4$
- $P(Gender = M|Wealthy = Y) = 1/4$
- $P(HrsWorked > 40.5|Wealthy = Y) = 1/4$
- $P(HrsWorked < 40.5|Wealthy = Y) = 3/4$
- $P(Gender = F|Wealthy = N) = 1/3$
- $P(Gender = M|Wealthy = N) = 2/3$
- $P(HrsWorked > 40.5|Wealthy = N) = 2/3$
- $P(HrsWorked < 40.5|Wealthy = N) = 1/3$

Based on the above, please use Classify a new instance with Naive Bayes algorithm (Gender = F, HrsWorked = 44).   □

Answer Because

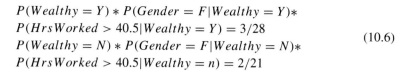

$$P(Wealthy = Y) * P(Gender = F|Wealthy = Y)*$$
$$P(HrsWorked > 40.5|Wealthy = Y) = 3/28$$
$$P(Wealthy = N) * P(Gender = F|Wealthy = N)*$$
$$P(HrsWorked > 40.5|Wealthy = n) = 2/21$$

(10.6)

we have that it will be predicted as $Wealthy = Y$ according to Eq. (8.11).   □

Question 11 Implement the perceptron learning algorithm in Sect. 10.1, and compare the obtained binary classifier with the one obtained from logistic regression.

□

Question 12 Understand how learning rate in the perceptron learning algorithm affects the learning results. Draw a curve to exhibit the change of accuracy with respect to the learning rate.   □

Question 13 Use the perceptron learning algorithm to work with the XOR dataset in Example 10.3, and check the accuracy.   □

Question 14 Assuming that all weights in Fig. 10.5, i.e., those numbers 1 and -1, need to be learned. Can you adapt the perceptron learning algorithm to learn the weights?   □

Question 15 Understand the equivalence of applying matrix expression to compute the outputs for a dataset and the computation of outputs for individual inputs.   □

Part III
Safety Solutions

After discovering safety vulnerabilities through algorithms in the previous part, it is a natural next step to consider safety solutions. In this part, we consider two major approaches: verification and enhancement. The relation between attack, verification, and enhancement is given in Fig. 1. Once safety attacks identify safety vulnerabilities of certain safety properties, the dedicated enhancement will be applied to improve the machine learning models before passing over to the safety verification, which in turn determines whether the safety properties hold. Once an affirmative answer is reported, we conclude that the enhanced model is safe with respect to the property. Otherwise, we will repeat the above process.

Verification is a collection of techniques that, given a model (e.g., a trained machine learning model) and a property, automatically determine whether a property holds on the model. Unlike safety attacks, the verification algorithm can conclude the existence or non-existence of safety vulnerabilities with mathematical proof. Therefore, it contributes as a key step in software development to ensure that the software achieves its design specification/requirement. Currently, verification for machine learning is still in its infancy, and we will focus on a recent surge in the verification of robustness property over the feedforward neural network.

Enhancement is usually specific with respect to the property under consideration. We discuss two typical enhancements in Chap. 12, for robustness and privacy, respectively. Robustness enhancement is conducted through adversarial training, which considers adversarial examples during the training process. On the other hand, privacy enhancement is conducted through randomisation, by adding noises to either the training or the inference.

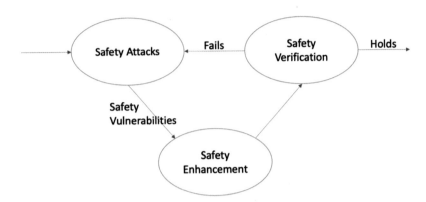

Fig. 1 Interactions of safety attacks (Part II) with two safety solutions to be discussed in this Part

Chapter 11
Verification of Deep Learning

Successful experience from industrial software engineering, which produced software that is currently applied in safety-critical applications, such as automotive and avionic applications, suggests that, to develop high-quality and low-cost software in a limited production time, a software development life cycle (SDLC) process is required. As illustrated in Fig. 11.1 about a V-model for SDLC, verification is a key process throughout the development lifecycle. This chapter will discuss a specific verification task, i.e., given a trained machine learning model and a robustness property, it is to determine whether the robustness property holds on the model. After the definition of robustness property in Sect. 11.1, we will present representative examples for two categories of verification algorithms in Sects. 11.2 and 11.3, respectively.

Existing verification algorithms can be roughly categorised into exhaustive search [70], constraint-solving based methods [82], abstract interpretation based methods [51, 98], global optimisation [145, 146], game-based methods [178, 181], and symbolic interval analysis [98, 99, 186]. The readers are referred to [68] for a survey. Moreover, in addition to the pixel perturbations measured with norm distances, we are also looking into real-world perturbations such as geometric and spatial perturbations [172]. This chapter presents a few typical verification algorithms. The first algorithm of this chapter (Sect. 11.2) reduces the verification to a constraint solving problem, which can then be solved with an off-the-shelf solver. The algorithm is white-box, and the reduction needs to consider the internal architecture of the neural networks. The complexity is NP-complete with respect to the combined number of hidden neurons and input features. On the other hand, the second algorithm (Sect. 11.3), as some others [145, 146, 178, 181, 172], is black-box, i.e., they do not rely on the internal architecture of the neural networks. Theoretically, this brings a significant advantage that the computational complexity of the verification problem is NP-complete with respect to the number of input features. While the complexity class does not change, the number of hidden neurons can be an unlimited number of times more than that of input features, due to the

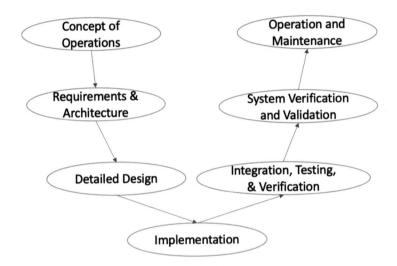

Fig. 11.1 An illustrative V-model for software development

current trend of deep learning on training deeper and larger networks. Also, black-box verification means that we can work with neural networks of any scale and structure.

11.1 Robustness Properties for Verification

A (deep and feedforward) neural network, or neural network, can be defined as a tuple $\mathcal{N} = (\mathbb{S}, \mathbb{T}, \Phi)$, where $\mathbb{S} = \{\mathbb{S}_k \mid k \in \{1..K\}\}$ is a set of layers, $\mathbb{T} \subseteq \mathbb{S} \times \mathbb{S}$ is a set of connections between layers and $\Phi = \{\phi_k \mid k \in \{2..K\}\}$ is a set of functions, one for each non-input layer. In a neural network, $\mathbb{S}_1$ is the *input* layer, $\mathbb{S}_K$ is the *output* layer, and layers other than input and output layers are called *hidden layers*. Each layer $\mathbb{S}_k$ consists of s_k *neurons* (or nodes). The l-th node of layer k is denoted by $n_{k,l}$.

Each node $n_{k,l}$ for $2 \leq k \leq K$ and $1 \leq l \leq s_k$ is associated with two variables $u_{k,l}$ and $v_{k,l}$, to record its values before and after an activation function, respectively. The Rectified Linear Unit (ReLU) [118] is one of the most popular activation functions for neural networks, according to which the *activation value* of each node of hidden layers is defined as

$$v_{k,l} = ReLU(u_{k,l}) = \begin{cases} u_{k,l} & \text{if } u_{k,l} \geq 0 \\ 0 & \text{otherwise} \end{cases} \tag{11.1}$$

Each input node $n_{1,l}$ for $1 \leq l \leq s_1$ is associated with a variable $v_{1,l}$ and each output node $n_{K,l}$ for $1 \leq l \leq s_K$ is associated with a variable $u_{K,l}$, because no activation function is applied on them. Other popular activation functions beside ReLU include: Sigmoid, Tanh, and Softmax.

Except for the nodes at the input layer, every node is connected to nodes in the preceding layer by pre-trained parameters such that for all k and l with $2 \leq k \leq K$ and $1 \leq l \leq s_k$

$$u_{k,l} = b_{k,l} + \sum_{1 \leq h \leq s_{k-1}} w_{k-1,h,l} \cdot v_{k-1,h} \tag{11.2}$$

where $w_{k-1,h,l}$ is the weight for the connection between $n_{k-1,h}$ (i.e., the h-th node of layer $k-1$) and $n_{k,l}$ (i.e., the l-th node of layer k), and $b_{k,l}$ the so-called *bias* for node $n_{k,l}$. We note that this definition can express both fully-connected functions and convolutional functions.[1] The function ϕ_k is the composition of Eqs. (11.1) and (11.2) by having $u_{k,l}$ for $1 \leq l \leq s_k$ as the intermediate variables. Owing to the use of the ReLU as in (11.1), the behavior of a neural network is highly non-linear.

Let $\mathbb{R}$ be the set of real numbers. We let $\mathcal{D}_k = \mathbb{R}^{s_k}$ be the vector space associated with layer $\mathbb{S}_k$, one dimension for each variable $v_{k,l}$. Notably, every point $\mathbf{x} \in \mathcal{D}_1$ is an input. Without loss of generality, the dimensions of an input are normalised as real values in $[0, 1]$, i.e., $\mathcal{D}_1 = [0, 1]^{s_1}$. A neural network $\mathcal{N}$ can alternatively be expressed as a function $f : \mathcal{D}_1 \to \mathcal{D}_K$ such that

$$f(\mathbf{x}) = \phi_K(\phi_{K-1}(\ldots \phi_2(\mathbf{x}))) \tag{11.3}$$

Finally, for any input, the neural network $\mathcal{N}$ assigns a *label*, that is, the index of the node of output layer with the largest value:

$$label = \mathrm{argmax}_{1 \leq l \leq s_K} u_{K,l} \tag{11.4}$$

Moreover, we let $C = \{1..s_K\}$ be the set of labels.

Example 11.1 Figure 11.2 is a simple neural network with four layers. The input space is $\mathcal{D}_1 = [0, 1]^2$, the two hidden vector spaces are $\mathcal{D}_2 = \mathcal{D}_3 = \mathbb{R}^3$, and the set of labels is $C = \{1, 2\}$.

Given one particular input $\mathbf{x}$, the neural network $\mathcal{N}$ is *instantiated* and we use $\mathcal{N}[\mathbf{x}]$ to denote this instance of the network. In $\mathcal{N}[\mathbf{x}]$, for each node $n_{k,l}$, the values of the variables $u_{k,l}$ and $v_{k,l}$ are fixed and denoted as $u_{k,l}[\mathbf{x}]$ and $v_{k,l}[\mathbf{x}]$, respectively. Thus, the activation or deactivation of each ReLU operation in the network is

---

[1] Many of the surveyed techniques can work with other types of functional layers such as max-pooling, batch-normalisation, etc. Here for simplicity, we omit their expressions.

Fig. 11.2 A simple neural network

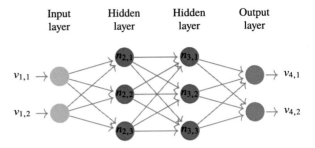

similarly determined. We define

$$sign_{\mathcal{N}}(n_{k,l}, \mathbf{x}) = \begin{cases} +1 & \text{if } u_{k,l}[\mathbf{x}] = v_{k,l}[\mathbf{x}] \\ -1 & \text{otherwise} \end{cases} \tag{11.5}$$

The subscript $\mathcal{N}$ will be omitted when clear from the context. The classification label of x is denoted as $\mathcal{N}[\mathbf{x}].label$.

Example 11.2 Let $\mathcal{N}$ be a neural network whose architecture is given in Fig. 11.2. Assume that the weights for the first three layers are as follows:

$$\mathbf{W}_1 = \begin{bmatrix} 4 & 0 & -1 \\ 1 & -2 & 1 \end{bmatrix}, \quad \mathbf{W}_2 = \begin{bmatrix} 2 & 3 & -1 \\ -7 & 6 & 4 \\ 1 & -5 & 9 \end{bmatrix}$$

and that all biases are 0. When given an input $\mathbf{x} = [0, 1]$, we get $sign(n_{2,1}, \mathbf{x}) = +1$, since $u_{2,1}[\mathbf{x}] = v_{2,1}[\mathbf{x}] = 1$, and $sign(n_{2,2}, \mathbf{x}) = -1$, since $u_{2,2}[\mathbf{x}] = -2 \neq 0 = v_{2,2}[\mathbf{x}]$.

11.1.1   Robustness as an Optimisation Problem

We have defined robustness in Sect. 3.3. For robustness verification, given an input $\mathbf{x}$, a d-neighbourhood $\eta(\mathbf{x}, L_p, d)$ (as in Definition B.1), and a neural network f, it is to check whether all inputs within the d-neighbourhood have the same label, i.e.,

$$\forall \mathbf{x}' : ||\mathbf{x} - \mathbf{x}'||_p < d \Rightarrow \hat{f}(\mathbf{x}) = \hat{f}(\mathbf{x}') \tag{11.6}$$

where $\hat{f}(\mathbf{x})$ returns the predictive label of $\mathbf{x}$ by f. Alternatively, this can be reduced to first finding the maximum safety radius δ such that

$$\delta \triangleq \min_{\mathbf{x}'} ||\mathbf{x} - \mathbf{x}'||_p \\ s.t. \ \hat{f}(\mathbf{x}) \neq \hat{f}(\mathbf{x}') \tag{11.7}$$

Then, it is to check whether $d \leq \delta$ as the verification result.

11.2 Reduction to Mixed Integer Linear Programming (MILP)

In this chapter, we present how we can reduce the verification problem (Eq. (11.7)) to the mixed integer linear programming (MILP) problems, so that it can be solved with the off-the-shelf MILP solvers. We will also consider the over-approximation of the problem so that it is able to be solved with linear programming. We will focus on the ReLU neural network (i.e., all activation functions are ReLU) and the robustness property.

11.2.1 Reduction to MILP

Let $\mathbf{x}_c$ be the original input whose label is $y_c = \hat{f}(\mathbf{x}_c)$. Recall from Chap. 11.1 that, we assume the network f has K layers. Then, Eq. (11.7) can be rewritten as

$$
\begin{aligned}
&\max_{\mathbf{x}} \; \|\mathbf{x} - \mathbf{x}_c\|_p \\
&s.t. \; \mathbf{x} = \mathbf{v}_1, \\
&\qquad \mathbf{up}_{i+1} = \mathbf{W}_i \mathbf{v}_i + \mathbf{b}_i, \qquad i = 1..K - 1 \\
&\qquad \mathbf{v}_{i+1} = ReLU(\mathbf{up}_{i+1}), \quad i = 1..K - 2 \\
&\qquad \mathbf{up}_K(y_c) - \mathbf{up}_K(y) \geq 0, \; y \in C
\end{aligned}
\tag{11.8}
$$

where $\mathbf{up}_i$, $\mathbf{v}_i$ denote the activation vector of layer i before and after the ReLU function, respectively. The first condition confirms to have $\mathbf{x}$ as the activation vector of the input layer. The second and third conditions implement the linear transformation and ReLU activation function of layer $i + 1$, respectively. The fourth condition requires that $\mathbf{x}$ has the label y_c. Specifically, the label is y_c if and and only if $\forall y \in C : \mathbf{up}_K(y_c) - \mathbf{up}_K(y) \geq 0$.

Considering that the ReLU function is non-linear, we introduce two methods of transforming the second and third conditions of Eq. (11.8) into MILP constraints, i.e., linear constraints with Boolean variables.

11.2.2 Method One for Layers

The first method requires one Binary variable for each neuron. Let $\mathbf{t}_{i+1}$ have value 0 or 1 in its entries and have the same dimension as $\mathbf{v}_{i+1}$, and M be a very large constant number that can be treated as ∞. We do not need $\mathbf{u}_{i+1}$. Specifically, we have the following MILP constraints for every layer $i = 1..K - 2$ to replace the second and third conditions of Eq. (11.8):

$$
\begin{aligned}
\mathbf{v}_{i+1} &\geq \mathbf{W}_i \mathbf{v}_i + \mathbf{b}_i, \\
\mathbf{v}_{i+1} &\leq \mathbf{W}_i \mathbf{v}_i + \mathbf{b}_i + M\mathbf{t}_{i+1}, \\
\mathbf{v}_{i+1} &\geq \mathbf{0}, \\
\mathbf{v}_{i+1} &\leq M(1 - \mathbf{t}_{i+1}),
\end{aligned}
\tag{11.9}
$$

To understand how it works, if $\mathbf{t}_{i+1} = \mathbf{0}$ then Eq. (11.9) can be simplified as $\mathbf{v}_{i+1} = \mathbf{W}_i \mathbf{v}_i + \mathbf{b}_i$ and $\mathbf{0} \leq \mathbf{v}_{i+1} \leq M$, which corresponds to the case of $\mathbf{up}_{i+1} \geq \mathbf{0}$. On the other hand, if $\mathbf{t}_{i+1} = 1$ then Eq. (11.9) is reduced to $\mathbf{v}_{i+1} = \mathbf{0}$, which corresponds to the case of $\mathbf{up}_{i+1} < \mathbf{0}$. These can be extended to work with the general case where the elements in $\mathbf{t}_{i+1}$ can be either 0 or 1. In such case, the inequalities in Eq. (11.9) can be dealt with in an element-wise way.

Further, if we have additional upper and lower bounds, $\mathbf{lo}_i$ and $\mathbf{up}_i$, for $\mathbf{v}_i$, then Eq. (11.8) can be rewritten into

$$
\begin{aligned}
\mathbf{v}_{i+1} &\geq \mathbf{W}_i \mathbf{v}_i + \mathbf{b}_i, \\
\mathbf{v}_{i+1} &\leq \mathbf{W}_i \mathbf{v}_i + \mathbf{b}_i - \mathbf{lo}_{i+1}\mathbf{t}_{i+1}, \\
\mathbf{v}_{i+1} &\geq \mathbf{0}, \\
\mathbf{v}_{i+1} &\leq \mathbf{up}_{i+1}(1 - \mathbf{t}_{i+1}),
\end{aligned}
\tag{11.10}
$$

Note that, the only differences with Eq. (11.9) are on the second and fourth conditions, where $\mathbf{lo}_{i+1}$ and $\mathbf{up}_{i+1}$ instead of the large number M are used. To understand how it works, if $\mathbf{t}_{i+1} = \mathbf{0}$ then Eq. (11.10) is reduced to $\mathbf{v}_{i+1} = \mathbf{W}_i \mathbf{v}_i + \mathbf{b}_i$ and $\mathbf{0} \leq \mathbf{v}_{i+1} \leq \mathbf{up}_{i+1}$, which corresponds to the case of $\mathbf{up}_{i+1} \geq \mathbf{0}$. On the other hand, if $\mathbf{t}_{i+1} = 1$ then Eq. (11.10) is reduced to $\mathbf{v}_{i+1} = \mathbf{0}$ and $\mathbf{W}_i \mathbf{v}_i + \mathbf{b}_i \leq \mathbf{v}_{i+1} \leq \mathbf{W}_i \mathbf{v}_i + \mathbf{b}_i - \mathbf{lo}_{i+1}$. Note that, $\mathbf{W}_i \mathbf{v}_i + \mathbf{b}_i - \mathbf{lo}_{i+1} > \mathbf{0}$. Therefore, this corresponds to the case of $\mathbf{up}_{i+1} < \mathbf{0}$. Similarly, Eq. (11.9) should be treated in the element-wise way. One of the approaches of computing lower and upper bounds can be seen from Sect. 11.2.6.

11.2.3   Method Two for Layers

Different from the first method, the second method focuses on the ReLU activation function. It requires both $\mathbf{u}_{i+1}$ and $\mathbf{v}_{i+1}$. Specifically, we have

$$
\begin{aligned}
\mathbf{u}_{i+1} &= \mathbf{W}_i \mathbf{v}_i + \mathbf{b}_i, \\
\mathbf{v}_{i+1} &\geq \mathbf{0} \\
\mathbf{v}_{i+1} &\geq \mathbf{u}_{i+1} \\
\mathbf{v}_{i+1} &\leq \mathbf{up}_{i+1} \odot \mathbf{t}_{i+1} \\
\mathbf{v}_{i+1} &\leq \mathbf{u}_{i+1} - \mathbf{lo}_{i+1} \odot (1 - \mathbf{t}_{i+1})
\end{aligned}
\tag{11.11}
$$

where $\odot$ is the element-wise multiplication. To understand how it works, if $t_{i+1} = 1$ then we have $0 \leq v_{i+1} = u_{i+1} \leq up_{i+1}$, which corresponds to the case of $up_{i+1} \geq 0$. On the other hand, if $t_{i+1} = 0$, then we have $u_{i+1} \leq v_{i+1} = 0 \leq u_{i+1} - lo_{i+1}$, which corresponds to the case of $up_{i+1} < 0$.

11.2.4 Optimisation Objective

For the optimisation objective $||x - x_c||_p$ in Eq. (11.8), different conversions are needed for different norm distance metric L_p.

For L_1 norm, we introduce auxiliary variables z, which bound the absolute value of $x - x_c$, i.e., let $z \leq x_c - x$ and $z \leq x - x_c$. Therefore, we have

$$
\begin{aligned}
\max_{x} \quad & \sum z \\
s.t. \quad & x = v_1, \\
& up_{i+1} = W_i v_i + b_i, \quad i = 1..K - 1 \\
& v_{i+1} = ReLU(up_{i+1}), \quad i = 1..K - 2 \\
& up_K(y_c) - up_K(y) \geq 0, \ y \in C \\
& z \leq x_c - x \\
& z \leq x - x_c
\end{aligned}
\tag{11.12}
$$

where $\sum z$ is the element-wise summation of the vector z. Note that $z < 0$. Therefore, the maximisation over $\sum z$ is to find the closest x (with respect to x_c and L_1 norm).

For L_∞ norm, we introduce a single auxiliary variables z_∞, which bound the L_∞ norm of $x - x_c$, i.e., let $z_\infty \leq x_c(i) - x(i)$ and $z_\infty \leq x(i) - x_c(i)$, for all $i \in [1..s_1]$. Therefore, we have

$$
\begin{aligned}
\max_{x} \quad & z_\infty \\
s.t. \quad & x = v_1, \\
& up_{i+1} = W_i v_i + b_i, \quad i = 1..K - 1 \\
& v_{i+1} = ReLU(up_{i+1}), \quad i = 1..K - 2 \\
& up_K(y_c) - up_K(y) \geq 0, \ y \in C \\
& z_\infty \leq x_c(i) - x(i), \quad i \in [1..s_1] \\
& z_\infty \leq x(i) - x_c(i), \quad i \in [1..s_1]
\end{aligned}
\tag{11.13}
$$

Note that $z_\infty \leq 0$. Therefore, the maximisation over z_∞ is to find the closest x (with respect to x_c and L_∞).

For L_2 norm, the objective becomes quadratic, and therefore we may have to use mixed integer quadratic programming (MIQP), without the need of auxiliary variable. That is,

$$
\begin{aligned}
\max_{\mathbf{x}} \quad & \sum_{i=1}^{s_1} (\mathbf{x}(i) - \mathbf{x}_c(i))^2 \\
s.t. \quad & \mathbf{x} = \mathbf{v}_1, \\
& \mathbf{up}_{i+1} = \mathbf{W}_i \mathbf{v}_i + \mathbf{b}_i, \qquad i = 1..K-1 \\
& \mathbf{v}_{i+1} = ReLU(\mathbf{up}_{i+1}), \quad i = 1..K-2 \\
& \mathbf{up}_K(y_c) - \mathbf{up}_K(y) \geq 0, \ y \in C
\end{aligned} \tag{11.14}
$$

11.2.5 Over-Approximation with Linear Programming

Equation (11.11) can be over-approximated with the below method (illustrated in Fig. 11.3) on the ReLU activation function.

Intuitively, when mapping from $\mathbf{up}_i(j)$ to $\mathbf{v}_i(j)$, instead of using the two lines from ReLU function, i.e., from $(\mathbf{lo}_i(j), 0)$ to $(0, 0)$ and from $(0, 0)$ to $(\mathbf{up}_i(j), \mathbf{up}_i(j))$, we can use the line from $(\mathbf{lo}_i(j), 0)$ to $(\mathbf{up}_i(j), \mathbf{up}_i(j))$ to over-approximate the value of $\mathbf{v}_i(j)$. With this idea, the last three conditions of Eq. (11.11) can be replaced with the below ones for every $j \in [1..s_i]$:

$$
\begin{cases}
\mathbf{v}_{i+1}(j) = 0 & \text{if } \mathbf{up}_{i+1}(j) \leq 0 \\
\mathbf{v}_{i+1}(j) = \mathbf{up}_{i+1}(j) & \text{if } \mathbf{lo}_{i+1}(j) \geq 0 \\
\mathbf{v}_{i+1}(j) \geq 0, \mathbf{v}_{i+1}(j) \geq \mathbf{up}_{i+1}(j), & \text{otherwise} \\
\mathbf{v}_{i+1}(j) \leq \dfrac{\mathbf{up}_{i+1}(j)(\mathbf{up}_{i+1}(j) - \mathbf{lo}_{i+1}(j))}{\mathbf{up}_{i+1}(j) - \mathbf{lo}_{i+1}(j)}
\end{cases} \tag{11.15}
$$

where the three conditions for the case when $\mathbf{up}_{i+1}(j) \geq 0$ and $\mathbf{lo}_{i+1}(j) \leq 0$ represent the triangle determined by the three lines such that the value of $\mathbf{up}_{i+1}(j)$ will be in the triangle. Therefore, according to the known upper and lower bounds $\mathbf{up}_{i+1}(j)$ and $\mathbf{lo}_{i+1}(j)$, one of the options in Eq. (11.15) will be chosen. Note that, all options are linear, without using any Binary variables.

Fig. 11.3 Relaxation of ReLU activation

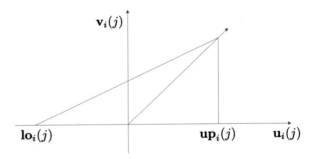

Based on the above discussion, if all ReLU activation functions are approximated with Eq. (11.15), the MILP problem becomes linear programming (LP), which can, in turn, be solved with an LP solver in linear time. We remark that, by approximating the MILP problem (whose computational complexity is NP-complete) with an LP problem (whose computational complexity is in P), solving the optimisation problem becomes tractable, so it is possible to work with a neural network of larger size. On the other hand, due to the over-approximation, when we are not able to affirm $d \leq \delta$ as the verification result (as we discussed after Eq. (11.7)), we are not able to draw a conclusion on the verification.

11.2.6   Computation of Lower and Upper Bounds Through Lipschitz Approximation

In this section, we consider how to approximate Lipschitz constant and how to utilise Lipschitz constant for the computation of lower and upper bounds $\mathbf{lo}_i$ and $\mathbf{up}_i$. Let

$$\mathbf{G}_i(\mathbf{x}) = \frac{\partial \mathbf{v}_i}{\partial \mathbf{x}} \qquad \text{for } i = 1..K \tag{11.16}$$

be the gradient vector of the hidden activation $\mathbf{v}_i$ over the input $\mathbf{x}$.

11.2.7   Approximation of Lipschitz Constant

We explain how to compute $\mathbf{G}_i(\mathbf{x})$ by utilising the structural information of the neural network. Actually, we will compute both its lower bound $\underline{\mathbf{G}}_i \in \mathbb{R}^{s_i \times s_1}$ and its upper bound $\overline{\mathbf{G}}_i \in \mathbb{R}^{s_i \times s_1}$. Before proceeding, we define a few notations and operators. Let $[a]_+ = \max\{a, 0\}$ and $[a]_- = \min\{a, 0\}$. For for a matrix $\mathbf{F}$, $[\mathbf{F}]_+$ and $[\mathbf{F}]_-$ are for element-wise max and min. Moreover, we define

$$\mathbf{W} \otimes [\mathbf{lo}, \mathbf{up}] = [[\mathbf{W}]_+ \times \mathbf{lo} + [\mathbf{W}]_- \times \mathbf{up}, [\mathbf{W}]_+ \times \mathbf{up} + [\mathbf{W}]_- \times \mathbf{lo}] \tag{11.17}$$

Actually, we can utilise the chain rule and the sequential structure of the neural network to have the following equation:

$$\mathbf{G}_i(\mathbf{x}) = \frac{\partial \mathbf{v}_i}{\partial \mathbf{x}} = \frac{\partial \mathbf{v}_i}{\partial \mathbf{u}_i} \frac{\partial \mathbf{u}_i}{\partial \mathbf{v}_{i-1}} \cdots \frac{\partial \mathbf{v}_2}{\partial \mathbf{u}_2} \frac{\partial \mathbf{u}_2}{\partial \mathbf{v}_1} \tag{11.18}$$

which suggests that the gradient $\mathbf{G}_i(\mathbf{x})$ can be computed repeatedly over the layers. Note that, $\frac{\partial \mathbf{u}_i}{\partial \mathbf{v}_{i-1}} = \mathbf{W}_{i-1}$, where $\mathbf{W}_{i-1}$ is the weight matrix of layer $i-1$. Let

$\nabla \sigma_i = \dfrac{\partial \mathbf{v}_i}{\partial \mathbf{u}_i} \in \mathbb{R}^{s_i \times s_i}$, we have

$$\mathbf{G}_i(\mathbf{x}) = \frac{\partial \mathbf{v}_i}{\partial \mathbf{x}} = \nabla \sigma_i \mathbf{W}_{i-1} \ldots \mathbf{W}_2 \nabla \sigma_2 \mathbf{W}_1 \qquad (11.19)$$

Let $\underline{\Lambda}_i, \overline{\Lambda}_i \in \mathbb{R}^{s_i \times s_i}$ be the diagonal matrices denoting the lower and upper bound of $\nabla \sigma_i$, respectively. For ReLU activation function, we have

$$\mathbf{1} \geq \overline{\Lambda}_i \geq \underline{\Lambda}_i \geq \mathbf{0} \qquad (11.20)$$

To enable a computation of the lower and upper bounds, we use an iterative process as follows. Let $\mathbf{F}_{i+1} = \mathbf{W}_i \mathbf{G}_i$ (i.e., $= \dfrac{\partial \mathbf{u}_{i+1}}{\partial \mathbf{x}}$), we have

$$\mathbf{G}_i = \nabla \sigma_i \mathbf{F}_i = \nabla \sigma_i \mathbf{W}_{i-1} \mathbf{G}_{i-1} \qquad (11.21)$$

The computation proceeds as follows. Initially, we have

$$\underline{\mathbf{G}}_0 = \overline{\mathbf{G}}_0 = \mathbf{I} \qquad (11.22)$$

where $\mathbf{I}$ is the identify matrix. Then, given $\underline{\mathbf{G}}_{i-1}$ and $\overline{\mathbf{G}}_{i-1}$, we have

$$\begin{aligned}
[\underline{\mathbf{F}}_i, \overline{\mathbf{F}}_i] &= \mathbf{W}_i \otimes [\underline{\mathbf{G}}_{i-1}, \overline{\mathbf{G}}_{i-1}] \\
\overline{\mathbf{G}}_i &= \max\{\underline{\Lambda}_i \overline{\mathbf{F}}_i, \overline{\Lambda}_i \overline{\mathbf{F}}_i\} = \overline{\Lambda}_i [\overline{\mathbf{F}}_i]_+ + \underline{\Lambda}_i [\overline{\mathbf{F}}_i]_- \\
\underline{\mathbf{G}}_i &= \min\{\underline{\Lambda}_i \underline{\mathbf{F}}_i, \overline{\Lambda}_i \underline{\mathbf{F}}_i\} = \underline{\Lambda}_i [\underline{\mathbf{F}}_i]_+ + \overline{\Lambda}_i [\underline{\mathbf{F}}_i]_-
\end{aligned} \qquad (11.23)$$

Finally, given an input region, which can be either the entire input domain or a d-neighbourhood, if we know $\overline{\Lambda}_i$ and $\underline{\Lambda}_i$, we can compute $[\underline{\mathbf{G}}_i, \overline{\mathbf{G}}_i]$ with respect to the input region. Therefore, this method can be used to compute either the global Lipschitz constant (when the input region is the entire input domain) or the local Lipschitz constant (when the input region is a d-neighbourhood).

11.2.8   Computation of Lower and Upper Bounds

Let $\eta(\mathbf{x}, L_p, d)$ be a d-neighbourhood centred around $\mathbf{x}$, we can first use the above method to compute $[\underline{\mathbf{G}}_i, \overline{\mathbf{G}}_i]$. Then, the upper bound and lower bounds of the j-th dimension are as follows:

$$\begin{aligned}
\mathbf{up}_i(j) &= \mathbf{v}_i(j) + \overline{\mathbf{G}}_i(j) \max_{\mathbf{x}' \in \eta(\mathbf{x}, L_p, d)} |\mathbf{x}(j) - \mathbf{x}'(j)| \\
\mathbf{lo}_i(j) &= \mathbf{v}_i(j) - \underline{\mathbf{G}}_i(j) \max_{\mathbf{x}' \in \eta(\mathbf{x}, L_p, d)} |\mathbf{x}(j) - \mathbf{x}'(j)|
\end{aligned} \qquad (11.24)$$

11.3 Robustness Verification via Reachability Analysis

As discussed in previous chapters, concerns have been raised about the suitability of deep neural networks (DNNs), or systems with DNN components, for deployment in safety-critical applications. To this end, besides those aforementioned verification techniques, we can also study a generic reachability problem in which, for a given DNN, an input subspace and a function over the network's outputs, computes the upper and lower bounds over the values of the function. The function is generic, with the only requirement that it is Lipschitz continuous. We argue that this problem is fundamental for the certification of DNNs, as it can be instantiated into several key correctness problems, including adversarial example generation [165, 56], safety verification [70, 82, 146], and output range analysis [102, 39].

To certify a system, a certification approach needs to provide not only a result but also a guarantee over the result, such as the error bounds. Existing approaches for analysing DNNs with provable guarantees work by either reducing the problem to a constraint satisfaction problem that can be solved by MILP [102, 23, 16, 182], SAT [120] or SMT [82, 16] solvers, or applying search algorithms over discretised vector spaces [70, 178]. Even though they are able to achieve guarantees, they suffer from two major weaknesses. Firstly, their subjects of study are restricted. More specifically, they can only work with layers conducting linear transformations (such as convolutional and fully-connected layers) and simple non-linear transformations (such as ReLU). They cannot work with other important layers, such as the Sigmoid, Max pooling and Softmax layers that are widely used in state-of-the-art networks. Secondly, the scalability of the constraint-based approaches is significantly limited by both the capability of the solvers and the size of the network. However, state-of-the-art networks usually have millions or even billions of hidden neurons.

This chapter will introduce a novel approach to tackle the generic reachability problem, which does not suffer from the above weaknesses and provides provable guarantees in terms of the upper and lower bounds over the errors. The approach is inspired by recent advances made in the area of global optimisation [53, 57]. For the input subspace defined over a set of input dimensions, an adaptive nested optimisation algorithm is developed. The performance of this algorithm is not dependent on the size of the network, and it can therefore scale to work with large networks.

This algorithm assumes certain knowledge about the DNN. However, instead of directly translating the activation functions and their parameters (i.e., weights and bias) into linear constraints, it needs a Lipschitz constant of the network. For this, we show that several layers that cannot be directly translated into linear constraints are actually Lipschitz continuous, and we can compute a Lipschitz constant by analysing the activation functions and their parameters. This method is implemented as a software tool DeepGO.[2]

---

[2] It is available on https://github.com/trustAI/DeepGO.

11.3.1 Lipschitz Continuity of Deep Learning

This section shows that feed-forward DNNs are Lipschitz continuous. Let $f : \mathbb{R}^n \to \mathbb{R}^m$ be a N-layer network such that, for a given input $\mathbf{x} \in \mathbb{R}^n$, $f(\mathbf{x}) = \{c_1, c_2, \ldots, c_m\} \in \mathbb{R}^m$ represents the confidence values for m classification labels. Specifically, we have $f(\mathbf{x}) = f_N(f_{N-1}(\ldots f_1(\mathbf{x}; \mathbf{W}_1, \mathbf{b}_1); \mathbf{W}_2, \mathbf{b}_2); \ldots); \mathbf{W}_N, \mathbf{b}_N)$ where $\mathbf{W}_i$ and $\mathbf{b}_i$ for $i = 1, 2, \ldots, N$ are learnable parameters and $f_i(\mathbf{z}_{i-1}; \mathbf{W}_{i-1}, \mathbf{b}_{i-1})$ is the function mapping from the output of layer $i - 1$ to the output of layer i such that $\mathbf{z}_{i-1}$ is the output of layer $i - 1$. Without loss of generality, we normalise the input $\mathbf{x} \in [0, 1]^n$. The output $f(\mathbf{x})$ is usually normalised to be in $[0, 1]^m$ with a Softmax layer.

Definition 11.1 (Lipschitz Continuity) Given two metric spaces $(\mathbf{X}, d_{\mathbf{X}})$ and $(\mathbf{Y}, d_{\mathbf{Y}})$, where $d_{\mathbf{X}}$ and $d_{\mathbf{Y}}$ are the metrics on the sets $\mathbf{X}$ and $\mathbf{Y}$ respectively, a function $f : \mathbf{X} \to \mathbf{Y}$ is called *Lipschitz continuous* if there exists a real constant $K \geq 0$ such that, for all $\mathbf{x}_1, \mathbf{x}_2 \in \mathbf{X}$:

$$d_{\mathbf{Y}}(f(\mathbf{x}_1), f(\mathbf{x}_2)) \leq K d_{\mathbf{X}}(\mathbf{x}_1, \mathbf{x}_2). \tag{11.25}$$

K is called the *Lipschitz constant* for the function f. The smallest K is called *the Best Lipschitz constant*, denoted as K_{best}.

The work in [165] shows that deep neural networks with half-rectified layers (i.e., convolutional or fully connected layers with ReLU activation functions), max pooling and contrast-normalization layers are Lipschitz continuous. They prove that the upper bound of the Lipschitz constant can be estimated via the operator norm of learned parameters $\mathbf{W}$. Furthermore, other researchers theoretically demonstrate that the Softmax layer, Sigmoid and Hyperbolic tangent activation functions also satisfy the Lipschitz continuity, the details of the proof can be found in the work of [144].

11.3.2 Reachability Analysis of Deep Learning

In this section, we present the formulate the problem of confidence reachability of a neural network. Let $o : [0, 1]^m \to \mathbb{R}$ be a Lipschitz continuous function statistically evaluating the outputs of the network. Our problem is to find its upper and lower bounds given the set $\mathbf{X}'$ of inputs to the network. Because both the network f and the function o are Lipschitz continuous, all values between the upper and lower bounds have a corresponding input, i.e., are reachable.

Definition 11.2 (Reachability of Neural Network) Let $\mathbf{X}' \subseteq [0, 1]^n$ be an input subspace and $f : \mathbb{R}^n \to \mathbb{R}^m$ a network. The reachability of f over the function o

under an error tolerance $\epsilon \geq 0$ is a set $R(o, \mathbf{X}', \epsilon) = [l, u]$ such that

$$
\begin{aligned}
\inf_{\mathbf{x}' \in \mathbf{X}'} o(f(\mathbf{x}')) - \epsilon \leq l &\leq \inf_{\mathbf{x}' \in \mathbf{X}'} o(f(\mathbf{x}')) + \epsilon \\
\sup_{\mathbf{x}' \in \mathbf{X}'} o(f(\mathbf{x}')) - \epsilon \leq u &\leq \sup_{\mathbf{x}' \in \mathbf{X}'} o(f(\mathbf{x}')) + \epsilon.
\end{aligned}
\tag{11.26}
$$

We write $u(o, \mathbf{X}', \epsilon) = u$ and $l(o, \mathbf{X}', \epsilon) = l$ for the upper and lower bound, respectively. Then the reachability diameter is

$$
D(o, \mathbf{X}', \epsilon) = u(o, \mathbf{X}', \epsilon) - l(o, \mathbf{x}', \epsilon).
\tag{11.27}
$$

Assuming these notations, we may write $D(o, \mathbf{X}', \epsilon; f)$ if we need to explicitly refer to the network f.

In the following, we instantiate o with a few concrete functions, and show that several key verification problems for DNNs can be reduced to our reachability problem.

Definition 11.3 (Output Range Analysis) Given a class label $j \in [1, .., m]$, we let $o = \Pi_j$ such that $\Pi_j((c_1, \ldots, c_m)) = c_j$.

We write $c_j(\mathbf{x}) = \Pi_j(f(\mathbf{x}))$ for the network's confidence in classifying $\mathbf{x}$ as label j. Intuitively, output range [39] quantifies how a certain output of a deep neural network (i.e., classification probability of a certain label j) varies in response to a set of DNN inputs with an error tolerance ϵ. Output range analysis can be easily generalised to logit[3] range analysis.

We show that the safety verification problem [70] can be reduced to solving the reachability problem.

Definition 11.4 (Local Safety) A network f is safe with respect to an input $\mathbf{x}$ and an input subspace $\mathbf{X}' \subseteq [0, 1]^n$ with $\mathbf{x} \in \mathbf{X}'$, written as $S(f, \mathbf{x}, \mathbf{X}')$, if

$$
\forall \mathbf{x}' \in \mathbf{X}' : \arg\max_j c_j(\mathbf{x}') = \arg\max_j c_j(\mathbf{x})
\tag{11.28}
$$

We have the following reduction theorem.

Theorem 11.1 *A network f is safe with respect to $\mathbf{x}$ and $\mathbf{X}'$ s.t. $\mathbf{x} \in \mathbf{X}'$ if and only if $u(\oplus, \mathbf{X}', \epsilon) \leq 0$, where $\oplus(c_1, \ldots, c_m) = \max_{i \in \{1..m\}} (\Pi_i(c_1, \ldots, c_m) - \Pi_j(c_1, \ldots, c_m))$ and $j = \arg\max_j c_j(\mathbf{x})$. The error bound of the safety decision problem by this reduction is 2ϵ.*

---

[3] Logit output is the output of the layer before the softmax layer. The study of logit outputs is conducted in, e.g., [39].

It is not hard to see that the adversarial example generation [165], which is to find an input $\mathbf{x}' \in \mathbf{X}'$ such that $\arg\max_j c_j(\mathbf{x}') \neq \arg\max_j c_j(\mathbf{x})$, is the dual problem of the safety problem.

Thus, by instantiating the function o, we can quantify a network's output/logit range and verify whether a network is robust or safe.

11.3.3   Confidence Reachability with Guarantees

Section 11.3.1 shows that a deep feedforward neural network is Lipschitz continuous regardless of its layer depth, activation functions and the number of neurons. Now, to solve the reachability problem, we need to find the *global* minimum and maximum values given an input subspace, assuming that we have a Lipschitz constant K for the function $o \cdot f$. In the following, we let $w = o \cdot f$ be the concatenated function. Without loss of generality, we assume the input space $\mathbf{X}'$ is a box-constraint (i.e., measured by L_∞-norm distance), which is clearly feasible since images are usually normalised into $[0, 1]^n$ before being fed into a neural network.

The computation of the minimum value is reduced to solving the following optimisation problem with guaranteed convergence to the global minimum (the maximisation problem can be similarly solved by minimising the negative objective function):

$$\min_{\mathbf{x}} w(\mathbf{x}), \quad s.t. \ \mathbf{x} \in [a, b]^n \tag{11.29}$$

However, the above optimisation is very challenging since $w(\mathbf{x})$ is a highly nonconvex function which cannot be guaranteed to reach the global minimum by regular optimisation schemes based on gradient descent. Inspired by an idea from optimisation, e.g., [136, 167], we design another continuous function $h(\mathbf{x}, \mathbf{y})$, which serves as a lower bound of the original function $w(\mathbf{x})$. Specifically, we need

$$\forall \mathbf{x}, \mathbf{y} \in [a, b]^n, \ h(\mathbf{x}, \mathbf{y}) \leq w(\mathbf{x}) \ \text{and} \ h(\mathbf{x}, \mathbf{x}) = w(\mathbf{x}) \tag{11.30}$$

Furthermore, for $i \geq 0$, we let $\mathcal{Y}_i = \{\mathbf{y}_0, \mathbf{y}_1, \ldots, \mathbf{y}_i\}$ be a finite set containing $i + 1$ points from the input space $[a, b]^n$, and let $\mathcal{Y}_i \subseteq \mathcal{Y}_k$ when $k > i$, then we can define a function $H(\mathbf{x}; \mathcal{Y}_i) = \max_{\mathbf{y} \in \mathcal{Y}_i} h(\mathbf{x}, \mathbf{y})$ which satisfies the following relation:

$$H(\mathbf{x}; \mathcal{Y}_i) < H(\mathbf{x}; \mathcal{Y}_k) \leq w(\mathbf{x}), \forall i < k \tag{11.31}$$

We use $l_i = \inf_{\mathbf{x} \in [a,b]^n} H(\mathbf{x}; \mathcal{Y}_i)$ to denote the minimum value of $H(\mathbf{x}; \mathcal{Y}_i)$ for $\mathbf{x} \in [a, b]^n$. Then we have

$$l_0 < l_1 < \ldots < l_{i-1} < l_i \leq \inf_{\mathbf{x} \in [a,b]^n} w(\mathbf{x}) \tag{11.32}$$

Similarly, we need a sequence of upper bounds u_i to have

$$l_0 < \ldots < l_i \leq \inf_{\mathbf{x} \in [a,b]^n} w(\mathbf{x}) \leq u_i < \ldots < u_0 \tag{11.33}$$

By Expression (11.33), we can have the following:

$$\lim_{i \to \infty} l_i = \min_{\mathbf{x} \in [a,b]^n} w(\mathbf{x}) \text{ and } \lim_{i \to \infty} (u_i - l_i) = 0 \tag{11.34}$$

Therefore, we can asymptotically approach the global minimum. Practically, we execute a finite number of iterations by using an error tolerance ϵ to control the termination. In the next section, we present the approach, which constructs a sequence of lower and upper bounds, and show that it can converge with an arbitrarily-small error bound.

11.3.3.1 One-Dimensional Case

We first introduce an algorithm which works over one dimension of the input, and therefore is able to handle the case of $x \in [a, b]$ in Eq. (11.29). The multi-dimensional optimisation algorithm will be discussed in the next section by repeatedly utilising the one-dimensional algorithm. We define the following lower-bound function.

$$h(x, y) = w(y) - K|x - y|$$
$$H(x; \mathcal{Y}_i) = \max_{y \in \mathcal{Y}_i} w(y) - K|x - y| \tag{11.35}$$

where $K > K_{best}$ is a Lipschitz constant of w and $H(x; \mathcal{Y}_i)$ intuitively represents the lower-bound saw-tooth function shown as Fig. 11.4. The set of points $\mathcal{Y}_i$ is constructed recursively. Assuming that, after $(i - 1)$-th iteration, we have $\mathcal{Y}_{i-1} = \{y_0, y_1, .., y_{i-1}\}$, whose elements are in ascending order, and sets

$$w(\mathcal{Y}_{i-1}) = \{w(y_0), w(y_1), .., w(y_{i-1})\}$$

$$\mathcal{L}_{i-1} = \{l_0, l_1, \ldots, l_{i-1}\}$$

$$\mathcal{U}_{i-1} = \{u_0, u_1, \ldots, u_{i-1}\}$$

$$\mathcal{Z}_{i-1} = \{z_1, \ldots, z_{i-1}\}$$

The elements in sets $w(\mathcal{Y}_{i-1})$, $\mathcal{L}_{i-1}$ and $\mathcal{U}_{i-1}$ have been defined earlier. The set $\mathcal{Z}_{i-1}$ records the smallest values z_k computed in an interval $[y_{k-1}, y_k]$.

In i-th iteration, we do the following sequentially:

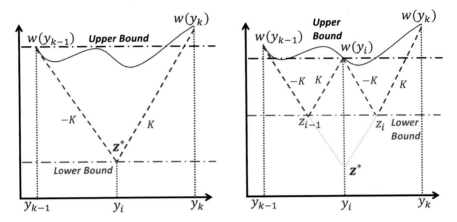

Fig. 11.4 A lower-bound function designed via Lipschitz constant

- Compute $y_i = \arg\inf_{x\in[a,b]} H(x; \mathcal{Y}_{i-1})$ as follows. Let $z^* = \min \mathcal{Z}_{i-1}$ and k be the index of the interval $[y_{k-1}, y_k]$ where z^* is computed. Then we let

$$y_i = \frac{y_{k-1} + y_k}{2} - \frac{w(y_k) - w(y_{k-1})}{2K} \qquad (11.36)$$

  and have that $y_i \in (y_{k-1}, y_k)$.
- Let $\mathcal{Y}_i = \mathcal{Y}_{i-1} \cup \{y_i\}$, then reorder $\mathcal{Y}_i$ in ascending order, and update $w(\mathcal{Y}_i) = w(\mathcal{Y}_{i-1}) \cup \{w(y_i)\}$.
- Calculate

$$z_{i-1} = \frac{w(y_i) + w(y_{k-1})}{2} - \frac{K(y_i - y_{k-1})}{2} \qquad (11.37)$$

$$z_i = \frac{w(y_k) + w(y_i)}{2} - \frac{K(y_k - y_i)}{2} \qquad (11.38)$$

  and update $\mathcal{Z}_i = (\mathcal{Z}_{i-1} \setminus \{z^*\}) \cup \{z_{i-1}, z_i\}$.
- Calculate the new lower bound $l_i = \inf_{x\in[a,b]} H(x; \mathcal{Y}_i)$ by letting $l_i = \min \mathcal{Z}_i$, and updating $\mathcal{L}_i = \mathcal{L}_{i-1} \cup \{l_i\}$.
- Calculate the new upper bound $u_i = \min_{y\in\mathcal{Y}_i} w(y)$ by letting $u_i = \min\{u_{i-1}, w(y_i)\}$.

We terminate the iteration whenever $|u_i - l_i| \le \epsilon$, and let the global minimum value be $y^* = \min_{x\in[a,b]} H(x; \mathcal{Y}_i)$ and the minimum objective function be $w^* = w(y^*)$.

Intuitively, as shown in Fig. 11.4, the key idea in this algorithm is to design a piecewise-linear lower bound function, which is guaranteed to be underneath the original function because of Lipschitz continuity. This lower bound function is refined iteration by iteration until the stopping criteria are satisfied. In each iteration, this algorithm is able to generate lower bounds by calculating the lowest point of the lower bound function; the upper bound is the lowest evaluation value of the original function so far.

11.3.3.2 Dynamically Improving the Lipschitz Constant

A Lipschitz constant closer to K_{best} can greatly improve the speed of convergence of the algorithm. We design a practical approach to dynamically update the current Lipschitz constant according to the information obtained from the previous iteration:

$$K = \eta \max_{j=1,\dots,i-1} \left| \frac{w(y_j) - w(y_{j-1})}{y_j - y_{j-1}} \right| \tag{11.39}$$

where $\eta > 1$. We emphasise that, with the optimisation proceeds, the more evaluations on the objective function we have, a more accurate estimation of the Lipschitz constant we can obtain, i.e.,

$$\lim_{i \to \infty} \max_{j=1,\dots,i-1} \eta \left| \frac{w(y_j) - w(y_{j-1})}{y_j - y_{j-1}} \right| = \eta \sup_{y \in [a,b]} \left| \frac{dw}{dy} \right| > K_{best}$$

The above analysis indicates that this dynamic estimation strategy can eventually approximate the true Lipschitz constant when iteration number i approximates to infinity.

11.3.3.3 Multi-Dimensional Case

The basic idea is to decompose a multi-dimensional optimisation problem into a sequence of nested one-dimensional sub-problems. Then the minima of those one-dimensional minimisation sub-problems are back-propagated into the original dimension and the final global minimum is obtained.

$$\min_{\mathbf{x} \in [a_i,b_i]^n} w(\mathbf{x}) = \min_{x_1 \in [a_1,b_1]} \dots \min_{x_n \in [a_n,b_n]} w(x_1, \dots, x_n) \tag{11.40}$$

We first introduce the k-th level sub-problem.

Definition 11.5 The k-th level optimisation sub-problem, written as $\phi_k(x_1, \dots, x_k)$, is defined as follows: for $1 \le k \le n - 1$,

$$\phi_k(x_1, \dots, x_k) = \min_{x_{k+1} \in [a_{k+1}, b_{k+1}]} \phi_{k+1}(x_1, \dots, x_k, x_{k+1})$$

and for $k = n$,

$$\phi_n(x_1, \ldots, x_n) = w(x_1, x_2, \ldots, x_n).$$

Combining Expression (11.40) and Definition 11.5, we have that

$$\min_{\mathbf{x} \in [a_i, b_i]^n} w(\mathbf{x}) = \min_{x_1 \in [a_1, b_1]} \phi_1(x_1)$$

which is actually a one-dimensional optimisation problem and therefore can be solved by the method in Sect. 11.3.3.1.

However, when evaluating the objective function $\phi_1(x_1)$ at $x_1 = a_1$, we need to project a_1 into the next one-dimensional sub-problem

$$\min_{x_2 \in [a_2, b_2]} \phi_2(a_1, x_2)$$

We recursively perform the projection until we reach the n-th level one-dimensional sub-problem,

$$\min_{x_n \in [a_n, b_n]} \phi_n(a_1, a_2, \ldots, a_{n-1}, x_n)$$

Once solved, we back-propagate objective function values to the first-level $\phi_1(a_1)$ and continue searching from this level until the error bound is reached.

The convergence analysis for both one-dimensional and multi-dimensional cases can be referred from [144]. Here we point out that, the proof in [144] indicates that the overall error bound of the nested scheme only increases linearly w.r.t. the bounds in the one-dimensional case.

11.3.4 A Running Numerical Example

In this section, we will use a numerical example to illustrate the fundamental differences between the reachability analysis method, DeepGO, constraint-solver based approach [16, 23, 44], and AI^2—a verification method based Abstract Interpretation (AI) [52].

Figure 11.5 shows the details of a neural network we used for investigation. This toy neural network has two input x_1 and x_2, two output y_1 and y_2. It contains one hidden layer with ReLU activation functions. We denote the ReLU activation by two parts: one part is r_1 and r_1 denoting the values before the activation, another part is h_1 and h_2 representing the values after activation. In this numerical example, we aim to solve the following reachability problem.

Problem 11.1 *Given a neural network and the box-constraints on its inputs, i.e., $x_1 \in [4, 6]$, $x_2 \in [4.5, 5]$, what is the output range of y_1?*

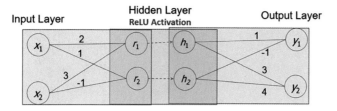

Fig. 11.5 Encoding the whole neural network layer by layer on a neural networks with one ReLU-based hidden layer

We will first show constraint-solver based solution, then reveal how AI^2 works and finally show how to use DeepGO to solve this problem.

11.3.4.1 Constraint-Solver Based Approach

As discussed in previous chapters, the majority of the solutions to solve Problem 11.1 are based on MILP or LP solvers, including SHERLOCK [39], Reluplex [82], Planet [44], MIP [23] and BaB [16], etc. We will reveal the basic idea of those works by using an LP-based solver as an example.

The first step is to encode the input and neural networks. As shown in Fig. 11.5, we can encode the whole neural network layer by layer.

$$
\begin{cases}
r_1 = 2x_1 + 3x_2 \\
r_2 = x_1 - x_2 \\
h_1 = \begin{cases} r_1, & \text{if } r_1 \geq 0. \\ 0, & \text{otherwise.} \end{cases} \\
h_2 = \begin{cases} r_2, & \text{if } r_2 \geq 0. \\ 0, & \text{otherwise.} \end{cases} \\
y_1 = h_1 - h_2 \\
y_2 = 3h_1 - 4h_2 \\
4 \leq x_1 \leq 6 \\
4.5 \leq x_2 \leq 5
\end{cases}
\tag{11.41}
$$

Then, to estimate the reachable interval of y_1, we also need to incorporate the target problem and formulate them into *four* linear programming problems based on the activation patterns of hidden neurons, as shown by the below equation.

$$\begin{cases} \max/\min & y_1 = x_1 + 4x_2 \\ \text{s.t.} & 2x_1 + 3x_2 \geq 0 \\ & x_1 - x_2 \geq 0 \\ & 4 \leq x_1 \leq 6 \\ & 4.5 \leq x_1 \leq 5 \end{cases} \cup \begin{cases} \max/\min & y_1 = 2x_1 + 3x_2 \\ \text{s.t.} & 2x_1 + 3x_2 \geq 0 \\ & x_1 - x_2 \leq 0 \\ & 4 \leq x_1 \leq 6 \\ & 4.5 \leq x_1 \leq 5 \end{cases}$$

$$\cup \begin{cases} \max/\min & y_1 = -x_1 + x_2 \\ \text{s.t.} & 2x_1 + 3x_2 \leq 0 \\ & x_1 - x_2 \geq 0 \\ & 4 \leq x_1 \leq 6 \\ & 4.5 \leq x_1 \leq 5 \end{cases} \cup \begin{cases} \max/\min & y_1 = 0 \\ \text{s.t.} & 2x_1 + 3x_2 \leq 0 \\ & x_1 - x_2 \leq 0 \\ & 4 \leq x_1 \leq 6 \\ & 4.5 \leq x_1 \leq 5 \end{cases}$$

$$(11.42)$$

By solving the above four linear programming problems using an LP solver, we can solve Problem 11.1 and calculate its reachable confidence interval [21.5, 26].

11.3.4.2   Abstract Interpretation Based Approach

A well-established verification work using Abstract Interpretation is AI^2 [52]. It phrases the problem of certifying neural networks in the classic abstract interpretation framework. The key ingredient in AI^2 is the layer-by-layer over-approximation of the neural network using zonotope-based abstract interpretation.

Figure 11.6 depicts the procedure of layer-by-layer zonotope abstract interpretation in AI^2 for solving Problem 11.1. AI^2 first adopts a zonotope to abstract all the inputs.

$$\mathbf{Z}_1 = \{(x_1, x_2) \mid x_1 = a_1 + 5; x_2 = 0.25a_2 + 4.75\} \qquad (11.43)$$

where $a_1 \in [-1, 1]$ and $a_2 \in [-1, 1]$.

Then it performs the zonotope abstract transformation based on the affine transformation from the input layer into the pre-activation layer. Please note that affine transformation is exact in zonotope-based abstraction.

$$\mathbf{Z}_2 = \{(r_1, r_2) \mid r_1 = 24.25 + 2a_1 + 0.75a_2; r_2 = 0.25 + a_1 - 0.25a_2\}$$

$$(11.44)$$

where $a_1 \in [-1, 1]$ and $a_2 \in [-1, 1]$.

Next, AI^2 considers the zonotope over-approximation on the first ReLU hidden neuron. Since $r_1 \geq 0$ always holds, zonohedron $\mathbf{Z}_2$ will transfer into the next layer without over-approximation loss. Thus we have $\mathbf{Z}_3 = \mathbf{Z}_2$. However, for the second

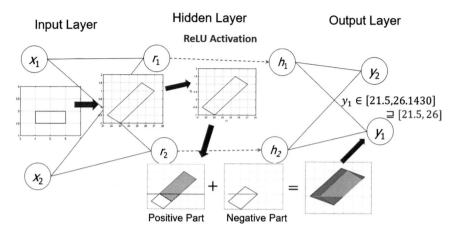

Fig. 11.6 Illustration of layer-by-layer over-approximation using zonotope abstract interpretation in AI^2

ReLU hidden neuron, the zonotope abstraction is complicated since r_2 is partially negative and partially positive, which requires two cases:

- Zonotope Over-approximation on Positive Part: As shown in Fig. 11.6, the positive part $\mathbf{Z}_3 \cap \{r_2 \geq 0\}$ is not a zonohedron. So we perform the zonotope over-approximation and get a new zonohedron:

$$\mathbf{Z}_{4,p} = \{(h_1, h_2) \mid h_1 = 24.75 + 1.5a_1 + 0.75a_2; h_2 = 0.5 + 0.75a_1 - 0.25a_2\}$$

- Zonotope Over-approximation on Negative Part: Similarly, we get a new zonotope

$$\mathbf{Z}_{4,n} = \{(h_1, h_2) \mid h_1 = 23.25 + 0.75a_1 + a_2; r_2 = 0\}$$

Then, as shown in Fig. 11.6, we perform a joint on two zonotopes $\mathbf{Z}_{4,p} \cup \mathbf{Z}_{4,n}$, and perform another zonotope over-approximation:

$$\mathbf{Z}_5 = \{(h_1, h_2) \mid h_1 = 24.3929 + 1.6429a_1 + 1.25a_2; h_2 = 0.5714 + 0.8214a_1 - 0.25a_2\} \tag{11.45}$$

where $a_1 \in [-1, 1]$ and $a_2 \in [-1, 1]$.

Finally, we can get the symbolic express on y_1 based on the affine transformation from after-activation layer to output layer:

$$y_1 = h1 - h2 = 23.8215 + 0.8215a_1 - 1.5a_2 \tag{11.46}$$

where $a_1 \in [-1, 1]$ and $a_2 \in [-1, 1]$. Thus, using AI^2, we can get the reachable output of y_1 is $[21.5, 26.1430] \supseteq [21.5, 26]$, which is an over-approximation of the actual reachable state of the neural network. As illustrated by Fig. 11.8, the yellow area is the actual information passed into the output layer, and the blue area is the over-approximation loss brought by the layer-by-layer zonotope abstractions.

11.3.4.3   Reachability Analysis by DeepGO

Now we show how DeepGO solve this reachability problem. As we can prove the target neural network in Fig. 11.4 is proved to be Lipschitz continuous [165, 144], to solve Problem 11.1, it can be reduced to solve the following minimisation problems.

$$\begin{cases} \min_{x_1, x_2} \quad y_1 = f(x_1, x_2) \\ \text{s.t.} \quad 4 \le x_1 \le 6 \\ \quad\quad 4.5 \le x_1 \le 5 \end{cases} \tag{11.47}$$

By solving the above two problems, we can get the reachable value of y_1. Figure 11.7 illustrates the optimisation procedure of DeepGO iteration by iteration.[4]

- Initialisation: It evaluates two ending points on x_1: 4 and 6, and gets the Upper Bound-1 and Lower Bound-1.
- Iterations: DeepGO then evaluates y_1 on the point of Lower Bound-1, and refines the lower bound to get Lower Bound-2 (described in Sect. 11.3.3.1). Similarly, we continue the optimisation iterations and get a series of lower bounds.
- Termination: After the gap between lower bound and upper bound is close enough, i.e., smaller than a positive number $\epsilon = 0.0001$, we stop and return the value of $y_1^* = 21.5$, which is the lowest value that the neural network can reach in Problem 11.1.

To get the largest reachable value, We replace objective function in Eq. (11.47) by $y_1 = -f(x_1, x_2)$ and perform the same procedure to get $y_1^{**} = 26$. By DeepGO, we can finally get the reachable interval of the neural network: $y_1 = [21.5, 26]$.

In summary, for Problem 11.1, as we can see from Fig. 11.8, DeepGO can calculate the reachable range of the neural network without an over-approximation error, but AI^2 brings a non-trivial error due to its over-approximation nature from the abstract interpretation. On the other hand, although LP/MILP based solution can also obtain an exact reachable range, DeepGO is much more efficient, especially for a neural network with a massive number of neurons.

---

[4] For visualisation, we just show the x_1 dimension.

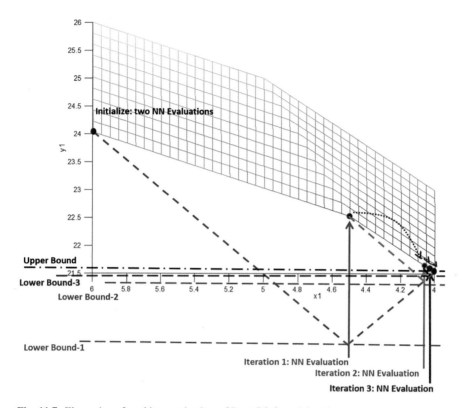

Fig. 11.7 Illustration of working mechanism of DeepGO for solving the reachability problem

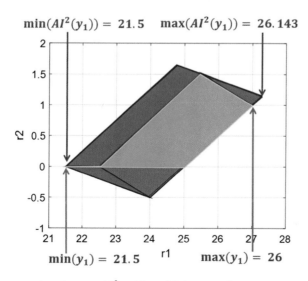

Fig. 11.8 The comparison between AI^2 and DeepGO in terms of over-approximation error

Chapter 12
Enhancement to Safety and Security of Deep Learning

Significant efforts from the research community have been spent on studying various methods to enhance either the training process or a trained model to mitigate the identified safety risks. In this chapter, we present three representative examples from three categories of techniques. They are designed to deal with different safety risks: robustness, generalisation, and privacy, respectively. The first category of techniques (Sect. 12.1), called adversarial training, is to improve the robustness. While there are many different approaches for the improvement of robustness, adversarial training is shown to be the most successful one. We will re-cap the typical min-max formalism of adversarial training, and then discuss some state-of-the-art training techniques. The second category of techniques (Sect. 12.2) is to improve the generalisation ability of a trained neural network. The technique presented, based on PAC Bayesian theory, utilises the weight correlation—a novel concept that measures the correlation between weights of the same layer—for training. The third category of techniques (Sect. 12.3) is to improve the privacy-related properties, such as membership inference, model stealing, and model inversion. They are based on differential privacy, and add random noise into either the training process or the inference stage.

12.1 Robustness Enhancement Through Min-Max Optimisation

Deep neural networks (DNNs) can be easily fooled to confidently make incorrect predictions by adding small and human-imperceptible perturbations to their input [56, 164, 179]. The study of adversarial defences, aiming to ultimately eliminate adversarial threat [87], has brought about significant advances in the past few years, with various techniques developed, including input denoising [7], adversarial detection [103], gradient regularisation [168], feature squeezing [185], defensive distillation [127] and adversarial training [104, 127]. Among various techniques, adversarial training is known to be the most effective one [6].

Adversarial training considers adversarial examples during the training. Consider a training dataset $D_{train} = \{(\mathbf{x}_i, y_i)\}_{i=1}^m$ with $m \in \mathbb{N}$ samples drawn from a distribution $\mathcal{D}$, where $\mathbf{x}_i \in \mathbb{R}^d$ is an example in the d-dimensional input space and y_i is its ground-truth label, adversarial training [104] updates the minimisation objective of the training scheme from the usual one as follows

$$J = \mathbb{E}_{(\mathbf{x},y)\sim\mathcal{D}} \mathcal{L}(\mathbf{x}; y; f_\theta) \tag{12.1}$$

to

$$J_{\text{adv}} = \mathbb{E}_{(\mathbf{x},y)\sim\mathcal{D}}\left[\max_{||\mathbf{x}'-\mathbf{x}||_p \le \epsilon} \mathcal{L}(\mathbf{x}'; y; f_\theta) \right], \tag{12.2}$$

where $\mathbf{x}'$ is the adversarial example within the ϵ-ball (bounded by an ℓ_p-norm) centred at clean example $\mathbf{x}$, $f_\theta(\cdot)$ is the DNN with parameter θ, and $\mathcal{L}(\cdot)$ is the standard classification loss (e.g., the cross-entropy loss). Therefore, adversarial training is formulated as a min-max optimisation problem.

On the opposite side of adversarial attacks [11, 165], researchers also show huge interest in designing various defence techniques, which are to either identify or reduce adversarial examples so that the decision of the DNN can be more robust. Until now, the developments of attack and defence techniques have been seen as an "arm-race". For example, most defences against attacks in the white-box setting, including [62, 110, 111, 127], have been demonstrated to be vulnerable to e.g., iterative optimisation-based attacks [17, 18].

12.1.1 Normal Adversarial Training

Adversarial training is one of the most notable defence methods, which was first proposed by Goodfellow et al. [56]. It can improve the robustness of DNNs against adversarial attacks by retraining the model on adversarial examples. Its basic idea can be expressed as below:

$$\theta^* = \arg\min_\theta \mathbb{E}_{(\mathbf{x},y)\sim\mathcal{D}} \mathcal{L}(\mathbf{x}; y; f_\theta). \tag{12.3}$$

This is improved in [104] by assuming that all neighbours within the ϵ-ball should have the same class label, i.e., local robustness. Technically, this is done by changing the optimisation problem by requiring that for a given ϵ-ball (represented as a d-Neighbourhood), to solve

$$\theta^* = \arg\min_\theta \mathbb{E}_{(\mathbf{x},y)\sim\mathcal{D}}\left[\max_{||\mathbf{x}'-\mathbf{x}||_p \le \epsilon} \mathcal{L}(\mathbf{x}'; y; f_\theta) \right]. \tag{12.4}$$

Intuitively, it is a min-max process, where each learning batch is conducted by first selecting the worst-case adversarial perturbation during the *inner maximisation*, and then adapting weights to reduce the loss by the adversarial perturbation in *outer minimisation*. [104] adopted Projected Gradient Descent (PGD) to approximately solve the inner maximisation problem. Please see the template code in Appendix D for the adversarial training.

Later on, to defeat the iterative attacks, [117] proposed to use a cascade adversarial method which can produce adversarial images in every mini-batch. Namely, at each batch, it performs a separate adversarial training by putting the adversarial images (produced in that batch) into the training dataset. Moreover, [168] introduces ensemble adversarial training, which augments training data with perturbations transferred from other models.

12.1.2 State-of-the-Art Adversarial Training Technology

The first category is to reduce Eq. (12.4) to an equivalent, or approximate, expression, which includes measuring the distance between $\mathbf{x}$ and $\mathbf{x}'$. For example, ALP [45, 79] enforces the similarity between $f_\theta(\mathbf{x})$ and $f_\theta(\mathbf{x}')$, the logits activations on unperturbed and adversarial versions of the same input $\mathbf{x}$. MMA [30] encourages every correctly classified instance $\mathbf{x}$ to leave a sufficiently large margin, i.e., the distance to the boundary, by maximising the size of the shortest successful perturbation. MART [174] observes the difference between misclassified and correctly classified examples in adversarial training, and suggests different loss functions for them. TRADES [193] analyses the robustness error and the clean error, and shows an upper bound and lower bound on the gap between robust error and clean error, which motivates adversarial training networks to optimise with

$$\mathcal{L}(\mathbf{x}; y; f_\theta) + \mathrm{KL}\big(f_\theta(\mathbf{x})\|f_\theta(\mathbf{x}')\big)/\lambda, \qquad (12.5)$$

where λ is the hyper-parameter to control the trade-off between clean accuracy and robust accuracy. It considers the KL-divergence of the activations of the output layer, i.e., $\mathrm{KL}(f_\theta(\mathbf{x})\|f_\theta(\mathbf{x}'))$, for every instance $\mathbf{x}$. The measurement over $\mathbf{x}$ and $\mathbf{x}'$ can be extended to consider a local distributional distance, i.e., the distance between the distributions within a norm ball of $\mathbf{x}$ and within a norm ball of $\mathbf{x}'$. For example, [201] forces the similarity between local distributions of an image and its adversarial example, [155] uses Wasserstein distance to measure the similarity of local distributions, and [33, 34, 35, 106, 124] optimise over distributions over a set of adversarial perturbations for a single image.

The second category is to pre-process the generated adversarial examples before training instead of directly using the adversarial examples generated by attack algorithms. Notable examples include label smoothing [22, 163], which, instead of considering the adversarial instances $(\mathbf{x}', y)$ for the "hard" label y, it consider $(\mathbf{x}', \tilde{\mathbf{y}})$, where $\tilde{\mathbf{y}}$ is a "soft" label represented as a weighted sum of the hard label and the uniform distribution. This idea is further exploited in [115], which empirically studies how label smoothing works. Based on these, AVMixup [95, 194] defines a virtual sample in the adversarial direction and extends the training distribution with soft labels via linear interpolation of the virtual sample and the clean sample. Specifically, it optimises

$$\mathcal{L}(\check{\mathbf{x}}, \check{\mathbf{y}}, f_\theta), \tag{12.6}$$

where $\check{\mathbf{x}} = \beta \mathbf{x} + (1-\beta)\gamma(\mathbf{x}' - \mathbf{x})$, $\check{\mathbf{y}} = \beta\phi(\mathbf{y}, \lambda_1) + (1-\beta)\phi(\mathbf{y}, \lambda_2)$, β is drawn from the Beta distribution for each single $\mathbf{x}_i$, γ is the hyper-parameter to control the scale of adversarial virtual vector, $\mathbf{y}$ is the one-hot vector of y, $\phi(\cdot)$ is the label smoothing function [163], and λ_1 and λ_2 are hyper-parameters to control the smoothing degree. Other than label smoothing, [192] generates adversarial examples by perturbing the local neighbourhood structure in an unsupervised fashion.

These two categories follow the min-max formalism and only adapt its components. AWP [180] adapts the inner maximisation to take one additional maximisation to find a weight perturbation based on the generated adversarial examples. The outer minimisation is then based on the perturbed weights [29] to minimise the loss induced by the adversarial examples. Specifically, it is to optimise the double-perturbation adversarial training problem

$$\max_{\mathbf{V} \in \mathcal{V}} \mathcal{L}(f_{\theta + \mathbf{V}}(\mathbf{x}')), \tag{12.7}$$

where $\mathcal{V}$ is a feasible region for the parameter perturbation $\mathbf{V}$.

In addition, [193] regularises the output of natural inputs and corresponding adversarial examples (generated by PGD attack) with the KL divergence. Xie and Yuille [184] explores the normalisation in adversarial training and studies the Mixture batch normalisation mechanism by using respective batch normalisation layers for natural data and adversarial examples. Cui et al. [26] proposes the Learnable Boundary Guided Adversarial Training (LBGAT) method, to improve robustness without losing much natural accuracy. Jin et al. [76] regularises second-order statistics of weights to improve robustness with the theoretical support from PAC-Bayesian adversarial bound.

12.2 Generalisation Enhancement Through PAC Bayesian Theory

Since the invention of the PAC-Bayesian framework [109], there have been a number of canonical works developed in the past decades, including the parametrisation of the PAC-Bayesian bound with Gaussian distribution [92, 93], the early works about generalisation error bounds for learning problems [54, 108, 130, 176] and the recent works about PAC-Bayes for machine learning models [2, 13, 40, 41, 43,

97, 133, 140, 166]. In the following, we will present a novelty method to improve DNN's generalisation performance through PAC-Bayesian theory.

Given a prior distribution over the parameter θ, which is selected before seeing a training dataset, a posterior distribution on θ will depend on both, the training dataset and a specific learning algorithm. The PAC-Bayesian framework [109] bounds the generalisation error with respect to the KL divergence [86] between the posterior and the prior distributions.

Consider a training dataset D_{train} with $m \in \mathbb{N}$ samples drawn from a distribution $\mathcal{D}$. Given a learning algorithm (e.g., a classifier) f_θ with prior and posterior distributions P and Q on the parameter θ respectively, for any $\delta > 0$, with probability $1 - \delta$ over the draw of training data, we have that [42, 109]

$$\mathbb{E}_{\theta \sim Q}[\mathcal{L}_{\mathcal{D}}(f_\theta)] \leq \mathbb{E}_{\theta \sim Q}[\mathcal{L}_{D_{train}}(f_\theta)] + \sqrt{\frac{KL(Q||P) + \log \frac{m}{\delta}}{2(m-1)}}, \qquad (12.8)$$

where $\mathbb{E}_{\theta \sim Q}[\mathcal{L}_{\mathcal{D}}(f_\theta)]$ is the expected loss on $\mathcal{D}$, $\mathbb{E}_{\theta \sim Q}[\mathcal{L}_{D_{train}}(f_\theta)]$ is the empirical loss on D_{train}, and their difference yields the generalisation error. Equation (12.8) outlines the role KL divergence plays in the upper bound of the generalisation error. In particular, a smaller KL term will help tighten the generalisation error bound. Assume that P and Q are Gaussian distributions with $P = \mathcal{N}(\mu_P, \Sigma_P)$ and $Q = \mathcal{N}(\mu_Q, \Sigma_Q)$, then the KL-term can be written as follows:

$$KL(\mathcal{N}(\mu_Q, \Sigma_Q)||\mathcal{N}(\mu_P, \Sigma_P))$$
$$= \frac{1}{2}\left[\text{tr}(\Sigma_P^{-1}\Sigma_Q) + (\mu_Q - \mu_P)^\top \Sigma_P^{-1}(\mu_Q - \mu_P) - k + \ln \frac{\det\Sigma_P}{\det\Sigma_Q}\right], \qquad (12.9)$$

where k is the number of parameters in θ. Below, [77] incorporates weight correlation into this framework to tighten the bound.

12.2.1 Weight Correlation in Fully Connected Neural Network (FCN)

Given weight matrix $w_l \in \mathbb{R}^{N_{l-1} \times N_l}$ of the l-th layer (N_l is the number of neurons of the l-th layer), the average weight correlation of FCN is defined as

$$\rho(N_l) = \frac{1}{N_l(N_l - 1)} \sum_{\substack{i,j=1 \\ i \neq j}}^{N_l} \frac{|w_{li}^T w_{lj}|}{||w_{li}||_2 ||w_{lj}||_2}, \qquad (12.10)$$

where w_{li} and w_{lj} are i-th and j-th column of the matrix w_l, corresponding to the i-th and j-th neuron at l-th layer, respectively. Intuitively, $\rho(w_l)$ is the average cosine similarity between weight vectors of any two neurons at l-th layer.

12.2.2   Weight Correlation in Convolutional Neural Network (CNN)

Given the filter tensor $w_l \in \mathbb{R}^{c \times c \times N_{l-1} \times N_l}$ of the l-th layer, where $c \times c$ is the size of the convolution kernel, $w_{li} \in \mathbb{R}^{c \times c \times N_{l-1}}$ and $w_{lj} \in \mathbb{R}^{c \times c \times N_{l-1}}$ are the i-th and j-th filter, respectively, of the filter tensor w_l. By reshaping w_{li} and w_{lj} into $w'_{li} \in \mathbb{R}^{c^2 \times N_{l-1}}$ and $w'_{lj} \in \mathbb{R}^{c^2 \times N_{l-1}}$, respectively, the weight correlation of CNN is defined as

$$\rho(w_l) = \frac{1}{N_l(N_l - 1)N_{l-1}} \sum_{\substack{i,j=1 \\ i \neq j}}^{N_l} \sum_{z=1}^{N_{l-1}} \frac{|w'^{T}_{li,z} w'_{lj,z}|}{||w'_{li,z}||_2 ||w'_{lj,z}||_2}, \tag{12.11}$$

where $w'_{li,z}$ and $w'_{lj,z}$ are the z-th column of w'_{li} and w'_{lj} respectively. Intuitively, $\rho(w_l)$ is defined as the cosine similarity between filter matrices (Fig. 12.1).

Considering the above $\rho(w_l)$, [77] introduces a correlation matrix $\Sigma_{\rho(w_l)} \in \mathbb{R}^{N_l \times N_l}$, with diagonal elements being 1, and off-diagonal ones all $\rho(w_l)$. Then, the posterior corvariance matrix can be represented as $\Sigma_{Q_{w_l}} = \Sigma_{\rho(w_l)} \otimes \sigma_l^2 I_{N_{l-1}}$, where $\otimes$ is Kronecker product. Let $g(w) = \sum g(w_l)$ where $g(w_l)$ defined by:

$$g(w_l) = -(N_l - 1)N_{l-1} \ln(1 - \rho(w_l)) - N_{l-1} \ln(1 + (N_l - 1)\rho(w_l)). \tag{12.12}$$

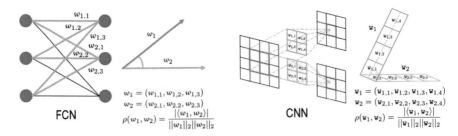

Fig. 12.1 **(FCN)** The weight correlation of any two neurons is the cosine similarity of the associated weight vectors. **(CNN)** The weight correlation of any two filters is the cosine similarity of the reshaped filter matrices

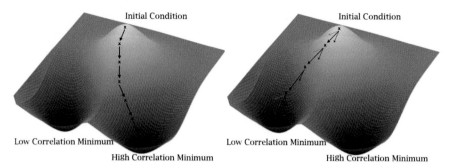

Fig. 12.2 **Left**: Normal gradient-based optimisers may find a local minimum with high correlation. **Right**: Weight correlation regularisation helps the optimiser to find a low correlation minimum, which is more likely to be a global minimum

Then the KL term w.r.t. the l-th layer can be given by

$$\mathrm{KL}(Q||P)_l = \frac{||\theta_l^F - \theta_l^0||_{\mathrm{Fr}}^2}{2\sigma_l^2} + g(w_l), \tag{12.13}$$

where θ^0 and θ^F refer to the value of parameters at initialisation and at the end of training, respectively. Further, when $\sigma_l^2 = \sigma^2$ for all l, we have $\mathrm{KL}(Q||P) = \sum_{l=1}^{L} \mathrm{KL}(Q||P)_l$. Given the above derivation, [77] concludes that the KL term in Eq. (12.13) is positively correlated to $g(w_l)$. Naturally, $g(w_l)$ can be considered as a regulariser term in the training function.

The training of a neural network is seen as a process of optimising over an objective function $J(\theta; \mathbf{x}, y)$. Jin et al. [77] adds a penalty term $g(w) = \sum_l g(w_l)$, which is a function of $\rho(w)$, to the objective function J, and denote the regularised objective function by $\tilde{J}$:

$$\tilde{J}(\theta; \mathbf{x}, y) = J(\theta; \mathbf{x}, y) + \alpha g(w), \tag{12.14}$$

where $\alpha \in [0, \infty)$ is a hyper-parameter that balances the relative contribution of the $g(w_l)$ penalty term. Figure 12.2 provides an illustrative diagram to show the utility of the regularisation. This regularisation is an effective and computationally efficient tool to enhance generalisation performance in practice [77], which could complement other commonly used regularisers such as weight decay and dropout.

12.3   Privacy Enhancement Through Differential Privacy

We have presented in Chap. 3 several security properties related to the leakage of privacy information, including membership inference and model inversion, and then discussed some algorithms that exploit different machine learning algorithms for privacy leakages, for example Chaps. 10.10 and 10.11 for deep learning algorithms. In this section, we consider how to enhance machine learning algorithms so that they may perform better in protecting the privacy information.

A straightforward idea for privacy protection is the naive anonymisation, by removing sensitive features such as names, addresses, and postcodes from the data. Unfortunately, an adversary can identify the user and uncover potentially sensitive information through auxiliary knowledge. For example, [119] shows that for machine learning model trained on Netflix Prize dataset, the attack can utilise auxiliary knowledge from the publicly available Internet Movie Database records.

While there are many different methods that have been proposed in the literature, we mainly focus on methods based on differential privacy (DP), because DP provides strong theoretical guarantees.

12.3.1 Differential Privacy

DP was originally developed when needing to publish aggregate information about a statistical database. It is required that the disclosure of private information of the records in the database is limited.

Let $\mathcal{A}$ be a randomised algorithm that generates an output according to a database. Let $O_{\mathcal{A}}$ be the set of possible outputs of $\mathcal{A}$, and we use $\mathcal{S}$ to range over $\mathcal{P}(O)$, i.e., the set of subsets of O.

Definition 12.1 The algorithm $\mathcal{A}$ is said to provide ϵ-differential privacy for ϵ a non-negative number, if for all datasets D_1 and D_2 that differ on a single element, we have

$$\forall \mathcal{S} \in \mathcal{P}(O_{\mathcal{A}}) : P(\mathcal{A}(D_1) \in \mathcal{S}) \leq e^{\varepsilon} \cdot P(\mathcal{A}(D_2) \in \mathcal{S}) \qquad (12.15)$$

Besides, we may also have (ϵ, δ)-differential privacy as follows.

Definition 12.2 The algorithm $\mathcal{A}$ is said to provide (ϵ, δ)-differential privacy for two non-negative numbers ϵ and δ, if for all datasets D_1 and D_2 that differ on a single element, we have

$$\forall \mathcal{S} \in \mathcal{P}(O_{\mathcal{A}}) : P(\mathcal{A}(D_1) \in \mathcal{S}) \leq \delta + e^{\varepsilon} \cdot P(\mathcal{A}(D_2) \in \mathcal{S}) \qquad (12.16)$$

Intuitively, δ represents the probability that the algorithm's output varies by more than a factor of e^{ϵ} when applied to a dataset D_1 and any one of its neighbours D_2. A smaller δ suggests a greater confidence and a smaller ϵ suggests a tighter standard of privacy protection. Actually, the smaller δ and ϵ are, the closer $P(\mathcal{A}(D_1))$ and $P(\mathcal{A}(D_2)) \in \mathcal{S})$ is, and therefore, the stronger privacy protection.

12.3.2   Private Algorithms for Training

While there are different proposals on integrating noises into training process to enhance the privacy protection (and implementing DP), we discuss a simple algorithm, DP-SGD [8], that lifts the stochastic gradient descent (SGD)—the training algorithm for deep learning—with noise. Typically, the SGD algorithm iterates over minibatchs, such that each minibatch is a small set of training examples. For every minibatch D, it updates the weight as follows:

$$\theta_{t+1} \leftarrow \theta_t + \eta \nabla_t \mathcal{L}(D; \theta_t) \tag{12.17}$$

where η is the learning rate, and $\mathcal{L}(D; \theta_t)$ is a loss function returning the average loss over the training instances in the minibatch D under the current weights θ_t. For DP-SGD, two modifications are made. First of all, the update is done on individual instances, instead of on a minibatch. Second, every gradient is added with a noise, i.e.,

$$\theta_{t+1} \leftarrow \theta_t + \eta (\nabla_t \mathcal{L}(\mathbf{x}; \theta_t) + b_t) \tag{12.18}$$

where $\mathbf{x}$ is a training sample randomly selected from D_{train} and b_t is sampled from a Gaussian noise $\mathcal{N}(0, \sigma^2)$ such that

$$\sigma = \sqrt{\frac{32L^2 n^2 \log(n/\delta) \log(1/\delta)}{\epsilon^2}} \tag{12.19}$$

This update is conducted for $|D_{train}|^2$ times. It is proved [8] that this algorithm is (ϵ, δ)-differential private.

12.3.3   Model Agnostic Private Learning

The above algorithm adapts the training algorithm for deep learning. In the following, we consider an algorithm, PATE [125], which does not rely on specific machine learning algorithm. Instead, the noise is added when aggregating results from multiple models trained over the partition of the dataset. Actually, the training dataset D_{train} is split into k datasets $D^1_{train}, \ldots, D^k_{train}$, each of which is used to train a model f^i for $i = 1..k$. Then, for each class $j \in C$, we let

$$n_j(\mathbf{x}) = |\{f^i(\mathbf{x}) = j \mid i = 1..k\}| \tag{12.20}$$

be the number of models that predict $\mathbf{x}$ as the label j. The privacy protected output is

$$f(\mathbf{x}) = \arg\max_{j \in C}(n_j(\mathbf{x}) + Lap(\frac{1}{\lambda})) \qquad (12.21)$$

where $Lap(b)$ is the Laplacian distribution with location 0 and scale b, and γ is a hyper-parameter. Intuitively, a large γ suggests a strong privacy guarantee, but may degrade the accuracy.

Exercises

Question 16 Please use an example to demonstrate your understanding of the dissimilarities of adversarial attacks and verification. □

Question 17 Please explain why L_p-norm distance metrics are important and how they were normally used in adversarial attacks for image classification models? (You can use one of the well-established attack methods as an example to facilitate the explanation) □

Question 18 Please explain why L_p-norm distance metrics are important and how they were normally used in adversarial attacks for image classification models? (You can use one of the well-established attack methods as an example to facilitate the explanation) □

Question 19 In robustness verification, some verification methods are sound, some are both sound and complete, please explain the soundness and completeness in verification. Could you please also name a few verification techniques/tools that are both sound and complete? □

Question 20 **Lipschitz Continuity**
     Given a neural network with one hidden layer with ReLU activation, shown as Fig. 12.3, please prove that the neural network is Lipschitz continuous. Please also calculate the Lipschitz constant of y_1 and y_2 w.r.t. x_1 and x_2. □

Question 21 **Reachability Problem**
     Given a neural network with one hidden layer of ReLU activation (shown as Fig. 12.3), assume $x_1 \in [3, 6.5]$ and $x_2 \in [2.5, 5.5]$, what is the output range of y_1 and y_2?

1. Please show how to solve the above reachability problem step by step using MILP/LP.
2. Please show how to solve the above reachability problem step by step using global optimisation (i.e., DeepGO). □

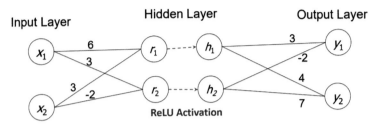

Fig. 12.3 A neural network with one hidden layer with ReLU activation

Question 22 Verification

Based on the solution of Question 6, show how to verify if $y_1 \leq y_2$ given $x_1 \in [3, 6.5]$ and $x_2 \in [2.5, 5.5]$? ☐

Question 23 Understand the basic idea of adversarial training, and implement an adversarial training algorithm with different step size, number of steps, epoch, to see which hyper-parameter setting can achieve the best balance between performance and running time. ☐

Question 24 Does adversarial training compromise the model's clean accuracy? If so, how to mitigate it? ☐

Question 25 Explore different assumptions on the distribution of random weights of DNNs, and understand which assumption is more reasonable in the PAC Bayesian theoretical framework. ☐

Question 26 Figure out other technologies to improve generalisation performance of DNNs. ☐

Part IV
Extended Safety Solutions

This part of the book will include several topics that are closely related to the solutions to machine learning safety but have not been covered in the previous part. It includes the consideration of other deep learning models (including deep reinforcement learning) and other safety analysis techniques (including testing techniques, reliability assessment, and assurance of lifecycle). These techniques should belong to the safety solutions, but unlike verification and enhancement in Part III, they are relatively new and we believe that additional efforts will be required before they become mature and are "upgraded" into safety solutions.

The first topic is to consider deep reinforcement learning (DRL), which has been applied as the control software for e.g., the motion planning of a mobile robot. We present several existing statistical evaluation methods that can roughly estimate the quality of a trained DRL policy. Then, we will consider probabilistic verification methods over an abstract model obtained from the example interactions of the DRL with the environment. The abstract model will vary according to different assumptions, e.g., whether the state-based robustness and/or whether the dynamics of robustness are considered. We also discuss sim-to-real challenge.

The second topic is about testing technique (or simulation-based verification) for neural networks, which has been widely discussed in the past years. Similar as software testing as opposed to software verification, testing technique for neural networks starts with the goal of striking a balance between the scalability and the cost for verification. In contrast to the exhaustiveness of verification, testing considers different definitions of coverage, based on which the test cases are generated.

The third topic is about reliability assessment, which aims to statistically evaluate high-level safety goals such as "the perception module will run without any misclassification for the next 1000 input instances with probability 99%". The estimation of high-level goals requires not only the statistical understanding of how the machine learning model is going to be used but also the support from safety verification or testing as safety evidence.

Finally, we suggest that the safety assurance of a machine learning model requires the consideration of its entire lifecycle, which includes four stages: data collection & preparation, model construction & training, verification & validation, and runtime enforcement. We discuss several subtopics that may appear in the assurance of machine learning lifecycle, including the assurance of data, sufficiency of training data, the optimised model construction and training, the adequacy of verification and validation, and the assured partnership of AI and humans.

Chapter 13
Deep Reinforcement Learning

This chapter consider another important application of machine learning to robotics, i.e., the utilisation of deep reinforcement learning agent for robot motion planning and control. We will first present some preliminaries on training DRL policy in robotics, followed by the discussion on the sample efficiency concerning the sufficiency of training (Sect. 13.3) and the introduction of several statistical methods for evaluation (Sect. 13.4). Afterwards, we will discuss how to formally express the properties (Sect. 13.5) and then focus on reusing the verification tools for convolutional neural network to work with deep reinforcement learning, by considering the verification of policy generalisation (Sect. 13.6), the verification of state-based policy robustness (Sect. 13.7), and the verification of temporal policy robustness (Sect. 13.8). In addition, we will discuss how to address the well-known Sim-to-Real challenge in robotics with the verification techniques (Sect. 13.9).

Unlike the convolutional neural network which relies on labelled data for supervised learning, reinforcement learning is a learning process in which the learning agent is trained through trial and error. Reinforcement learning is usually formalised as the finding of an optimal strategy in a Markov decision process (MDP). In other words, it is typical to model the interaction of an intelligent agent with its environment as an MDP, and then apply some algorithms (e.g., value iteration, policy iteration) to compute an optimal strategy for the intelligent agent. To utilise traditional MDP algorithm, it is needed to have several components well defined, including the states, the actions, the transition relation, and the reward function. However, for a real-world application, such components may not be easily defined, for example, the transition relation may not be definable as it can be impossible to have a complete definition of the environment. Moreover, some components such as the states may be of very high dimensional, which will make the traditional MDP algorithms fail to work due to the time and memory limitations. To deal with these problems, deep reinforcement learning (DRL) combines reinforcement learning and deep learning to enable our working with unstructured, high-dimensional data without manual engineering of the components. Another major difference

© The Author(s), under exclusive license to Springer Nature Singapore Pte Ltd. 2023   219
X. Huang et al., *Machine Learning Safety*, Artificial Intelligence: Foundations, Theory, and Algorithms, https://doi.org/10.1007/978-981-19-6814-3_13

from the convolutional neural network is the definition of safety properties. While for convolutional neural network the safety properties are closely related to the misclassification of individual input instances, for reinforcement learning the safety properties are more appropriate to be defined on the *sequential inputs*.

We remark that, in this chapter, we assume that the reward function is well defined, and it is based on this that we consider the safety properties. Other important factors related to the definition of rewards, such as the side effect and reward hacking [4], are not considered.

13.1   Interaction of Agent with Environment

We use discounted infinite-horizon MDP to model the interaction of an agent with the environment E. An MDP is a 5-tuple $\mathcal{M}^E = (\mathcal{S}, \mathcal{A}, \mathcal{P}, \mathcal{R}, \gamma)$, where $\mathcal{S}$ is the state space, $\mathcal{A}$ is the action space, $\mathcal{P}(\mathbf{x}'|\mathbf{x}, \mathbf{a})$ is a probabilistic transition, $\mathcal{R}(\mathbf{x}, \mathbf{a}) \in \mathbb{R}_{\geq 0}$ is a reward function, and $\gamma \in [0, 1)$ is a discount factor. We use $\mathbf{x}$ to range over the state space $\mathcal{S}$ because it not only is a state but also will later be used as input to a policy neural network. We consider DDPG [100, 162, 112] for a reinforcement learning algorithm, although there are many other deep reinforcement learning algorithms. DDPG returns a deterministic policy. A (deterministic) policy π includes a mapping $\mu : \mathcal{S} \rightarrow \mathcal{A}$ that maps from states to actions.

Based on $\mathcal{M}^E$, a policy π induces a trajectory distribution $\rho^{\pi, E}(\zeta)$ where

$$\zeta = (\mathbf{x}_0, \mathbf{a}_0, \mathbf{x}_1, \mathbf{a}_1, \ldots) \tag{13.1}$$

denotes a random trajectory. The state-action value function of π is defined as

$$Q^\pi(\mathbf{x}, \mathbf{a}) = \mathbb{E}_{\zeta \sim \rho^{\pi, E}}[\sum_{t=0}^{\infty} \gamma^t \mathcal{R}(\mathbf{x}_t, \mathbf{a}_t)] \tag{13.2}$$

and the state value function of π is

$$V^\pi(\mathbf{x}) = Q^\pi(\mathbf{x}, \pi(\mathbf{x})). \tag{13.3}$$

Example 13.1 We consider a reinforcement learning driven robot that navigates, and avoids collisions, in a complex environment where there are static and dynamic objects (or obstacles). The interaction of robot with the environment can be modelled as an MDP. At each time t, the robot has its observation of the laser sensors from the environment, namely state $\mathbf{x}_t$, i.e.,

$$\mathbf{x}_t = (o_t^1, o_t^2, \cdots, o_t^n)^T \tag{13.4}$$

where $o_t^1, o_t^2, \cdots, o_t^n$ are sensor signals at time t. The sensors can only observe partial information of the environment, e.g., by scanning the environment within a certain distance. For example, the observation range is within 3.15 m in Turtlebot Waffle Pi [141] for a distance sensor.

An action $\mathbf{a}_t \in \mathcal{A}$ consists of several decision variables. With the PID controller on the robot, we consider two action variables, representing line velocity and angle velocity, respectively, i.e.,

$$\mathbf{a}_t = (v_t^{line}, v_t^{angle})^T. \tag{13.5}$$

At each time t, the DRL policy outputs an action $\mathbf{a}_t$ from the action set $\mathcal{A}$.

The objective of the robot is to avoid the obstacles and reach a goal area. On every state $\mathbf{x}_t$, the sensory input o_t^i can be utilised to e.g., predict the distance to the obstacles and the goal area when they are close enough (within 3.15 m). To implement the objective, the environment may impose a reward function $\mathcal{R}$ on the states or the actions or both. A reward on the states can be e.g., with respect to the distance to obstacles, and a reward on the actions can be e.g., with respect to the acceleration in linear or angular speed.

13.2 Training a Reinforcement Agent

The model-free DDPG algorithm [100] is applied for the training of a DRL policy for the robot. Typically, the DRL policies are trained in a simulation environment before applied to the real world [24]. That is because of the unbearable costs of having real-world (negative) examples for training in real world [190]. The objective of the DDPG algorithm for an intelligent agent is to learn a policy π, which maximises the expectation of the best reward over time N:

$$\mathcal{J} = \mathbb{E}(\sum_{t=0}^{N-1} \gamma^t r_t | \mathbf{x}_t = \mathbf{x}_0) \tag{13.6}$$

where $\mathbf{x}_0$ is the initial state; r_t is the reward based on state-action pair at time slot t; γ is the discount factor which is applied to reduce the effect of future reward. To simplify the notation, we let $G_t = \sum_{t=0}^{N-1} \gamma^t r_t$.

The DDPG algorithm has two different neural networks, actor networks $\mu(\mathbf{x}_t | \theta^\mu)$ and critic networks $\nu(\mathbf{x}_t, \mathbf{a}_t | \theta^\nu)$, which are illustrated in Fig. 13.1. θ^μ and θ^ν are the weights of the actor and critic network, respectively. Due to the non-linearity of the neural networks, the DDPG algorithm can deal with the continues states and continues actions, which are more realistic to autonomous systems. Actor network is used to yield a deterministic action value $\mathbf{a}_t$, it takes the observation of environment as the inputs, and outputs the decided actions of the robot. Critic network is used to approximate Q value, which is use to determine whether the state-action pair

Fig. 13.1 Structure of DDPG

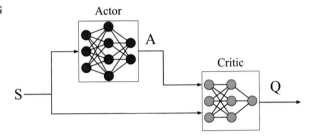

is good or not. It takes the observation from environment and the action value from actor networks, and then outputs the approximated $v(\mathbf{x}_t, \mathbf{a}_t)$ values. In the algorithm, the critic network is trained to minimise the loss function based on the stochastic gradient descent [100]:

$$\mathcal{L}(\theta^v) = \mathbb{E}[(y_t - v(\mathbf{x}_t, \mathbf{a}_t|\theta^v))^2] \tag{13.7}$$

where $y_t = r_t(\mathbf{x}_t, \mathbf{a}_t) + \gamma v(\mathbf{x}_{t+1}, \mu(\mathbf{x}_t|\theta^\mu)|\theta^v)$ is the approximated Q value based on current state $\mathbf{x}_t$ and previous parameters θ^μ, θ^v of two neural networks. The actor network is updated by the policy gradient with following equation [100]:

$$\nabla_{\theta^\mu} \mathcal{J}^{\theta^\mu} = \mathbb{E}[\nabla_{\mathbf{a}} v(\mathbf{x}, \mathbf{a}|\theta^v)|_{\mathbf{a}=\mu(\mathbf{x}|\theta^\mu)} \nabla_{\theta^\mu} \mu(\mathbf{x}|\theta^\mu)] \tag{13.8}$$

For the training, to breaking harmful correlations and learn from individual tuples for multiple times, an experience replay buffer is applied.

For DDPG, a learned policy π includes both actor network μ and critic network v. For some other DRL algorithms such as DQN, a learned policy π may include only a network μ.

13.3 Sufficiency of Training Data

Theoretically, it is shown in [175] that the sample complexity, i.e., the number of training-samples needed in order to successfully train a good model, is exponential with respect to either the input dimension or the finite horizon (of the sampled episodes). However, practical cases might not be as pessimistic as this looks like.

Empirically, we can adapt the learning curve idea (which plots the test accuracy with respect to the training set size) to this context, and plot the curve to show the achieved performance as the function of the number of episodes played during the learning. Actually, the number of episodes played during the learning is a concept similar as the size of training dataset for the supervised training of classifiers, and we can define the achieved performance as e.g., expected reward over a number of rollouts. By tracing the curve, we will be able to know when additional episodes will not make a significant change to the training result.

13.4   Statistical Evaluation Methods

Before introducing verification techniques which are usually of high computational complexity and hence are more suitable for safety-critical applications, we need some low-complexity evaluation methods to understand roughly how well a DRL algorithm works during training and how well a trained DRL policy works in an environment. Those model evaluation methods for general machine learning models (such as ROC and PR curves) can still be applicable, but in this section we will introduce a few DRL specific evaluation methods.

The below methods are all based on certain random variable X, and to track the trend and variance of X in the training or test phase.

13.4.1   Evaluation of DRL Training Algorithm

In the training phase, the random variable X can be e.g., per-epoch reward. Given a set of X's values, they form a distribution. It is useful to use the dispersion (also called variability, scatter, or spread) of this distribution to understand if the training process goes well. The measurement of the dispersion of a distribution can be with a few indicators [20] such as

- Interquartile Range (IQR). A distribution can be divided into 100 percentiles, denoted as $Q_1, \ldots, Q_{100}$, respectively. IQR can be defined as the difference between e.g., the 25th and 75th percentiles, i.e.,

$$\mathrm{IQR}(X) = Q_{75}(X) - Q_{25}(X) \qquad (13.9)$$

- Conditional Value at Risk (CVaR). CVaR concerns the worst cases, by taking a weighted average of the worst-case losses in the left-most tail of the distribution of possible outcomes, formally

$$\mathrm{CVaR}_\alpha(X) = \mathbb{E}[X \mid X \leq \mathrm{VaR}_\alpha(X)] \qquad (13.10)$$

where VaR_α (Value at Risk) is just the α-th quantile (e.g., quartile, percentile, or decile) of the distribution of X.

Then, we may consider the change of the above indicators over the training time in a single training run or across different training runs. For a single training run, we can split it into a number of sliding widows and then collect X's values from the sliding windows. When working with a set of different training runs, we can collect X's values from those runs.

13.4.2 Evaluation of DRL Model

Once there is a trained DRL policy, we can track its performance over a rollout by plotting the cumulative reward as a function of the number of steps. On such a plot, we may concern e.g., the slope, the minimum, and the zero-crossing (a point where the sign of the curve changes).

Moreover, we can also consider X as the performance over a set of rollouts, and use the IQR and CVaR to understand the dispersion of the distribution of X.

13.5 Safety Properties Through Probabilistic Computational Tree Logic

Probabilistic model checking [90] has been used to analyse quantitative properties of systems across a variety of application domains. It involves the construction of a probabilistic model, e.g., DTMC or MDP, that formally represents the behaviour of a system over time. The properties of interest are usually specified with, e.g., LTL or PCTL. Then, via model checkers, a systematic exploration and analysis is performed to check if a claimed property holds. In this section, we adopt DTMC and PCTL whose definitions are as follows.

Definition 13.1 (DTMC) Let AP be a set of atomic propositions. A DTMC is a tuple $(S, \mathbf{x}_0, \mathbf{P}, L)$, where S is a (finite) set of states, $\mathbf{x}_0 \in S$ is an initial state, $\mathbf{P} : S \times S \to [0, 1]$ is a probabilistic transition matrix such that $\sum_{s' \in S} \mathbf{P}(s, s') = 1$ for all $s \in S$, and $L : S \to 2^{AP}$ is a labelling function assigning each state with a set of atomic propositions.

Definition 13.2 (DTMC Reward Structure) A reward structure for DTMC $D = (S, \mathbf{x}_0, \mathbf{P}, L)$ is a tuple $r = (r_S, r_T)$ where $r_S : S \to \mathbb{R}_{\geq 0}$ is a state reward function and $r_T : S \times S \to \mathbb{R}_{\geq 0}$ is a transition reward function.

Definition 13.3 (PCTL) The syntax of PCTL is defined by *state formulae* ϕ, *path formulae* ψ and *reward formulae* μ.

$$\phi ::= true \mid ap \mid \phi \wedge \phi \mid \neg \phi \mid P_{\bowtie p}(\psi) \mid R^r_{\bowtie q}(\mu)$$

$$\psi ::= \phi \mid \phi \, U \, \phi$$

$$\mu ::= C^{\leq t} \mid \Diamond \phi$$

where $ap \in AP, p \in [0, 1], q \in \mathbb{R}_{\geq 0}, t \in \mathbb{N}, \bowtie \in \{<, \leq, >, \geq\}$ and r is a reward structure. The temporal operator is called "next", and U is called "until". We write $\Diamond \phi$ for $true \, U \, \phi$, and call it "eventually". Operator $C^{\leq t}$ is "bounded cumulative reward", expressing the reward accumulated over t steps. Formula

$R^r_{\bowtie q}(\lozenge \phi)$ expresses "reachability reward", the reward accumulated up until the first time a state satisfying ϕ.

Given $D = (S, \mathbf{x}_0, \mathbf{P}, L)$ and $r = (r_S, r_T)$, the satisfaction of state formula ϕ on a state $s \in S$ is defined as:

$$s \models true; \quad s \models ap \Leftrightarrow ap \in L(s); \quad s \models \neg \phi \Leftrightarrow s \not\models \phi;$$

$$s \models \phi_1 \wedge \phi_2 \Leftrightarrow s \models \phi_1 \text{ and } s \models \phi_2;$$

$$s \models \mathcal{P}_{\bowtie p}(\psi) \Leftrightarrow Pr(s \models \psi) \bowtie p;$$

$$s \models \mathcal{R}^r_{\bowtie q}(\mu) \Leftrightarrow \mathbb{E}[rew^r(\mu)] \bowtie q,$$

where $Pr(s \models \psi) \bowtie p$ concerns the probability of the set of paths that satisfy ψ and start in s. Given a path η, if write $\eta[i]$ for its i-th state and $\eta[0]$ the initial state, then

$$rew^r(C^{\leq t})(\eta) = \sum_{j=0}^{k-1}(r_S(\eta[j]) + r_T(\eta[j], \eta[j+1]))$$

$$rew^r(\lozenge \phi)(\eta) = \begin{cases} \infty & \forall j \in \mathbb{N}(\eta[j] \not\models \phi) \\ rew^r(C^{\leq ind(\eta,\phi)})(\eta) & \text{otherwise} \end{cases}$$

where $ind(\eta, \phi) = \min\{j \mid \eta[j] \models \phi\}$ denotes the index of the first occurrence of ϕ on path η. Moreover, the satisfaction relations for a path formula ψ on a path η is defined as:

$$\eta \models \phi \Leftrightarrow \eta[1] \models \phi$$

$$\eta \models \phi_1 U \phi_2 \Leftrightarrow \exists j \geq 0(\eta[j] \models \phi_2 \wedge \forall k < j(\eta[k] \models \phi_1))$$

Very often, it is of interest to know the actual probability that a path formula is satisfied, rather than just whether or not the probability meets a required threshold since this can provide a notion of margins as well as benchmarks for comparisons following later updates. So, the PCTL definition can be extended to allow *numerical queries* of the form $\mathcal{P}_{=?}(\psi)$ or $\mathcal{R}^r_{=?}(\psi)$ [90]. After formalising the system behaviors and properties in DTMC and PCTL respectively, automated tools have been developed to solve the verification problem, e.g., PRISM [89] and STORM [27]. We remark that, PCTL can be utilised to describe safety-related properties for e.g., the robot navigation example Example 13.1 as discussed in [32].

In the following sections, we will discuss several instantiations of the DTMC as $\mathcal{M}^E(\pi, \mathbf{x}_0)$ (Sect. 13.6), $\mathcal{M}^E(\pi, \mathbf{x}_0, C)$ (Sect. 13.7), $\mathcal{M}^E(\pi, C(\mathbf{x}_0))$ (Sect. 13.8), and $\mathcal{M}^{E_1 \times E_2}(\pi, (\mathbf{x}_0, \mathbf{x}_0))$ (Sect. 13.9), respectively.

13.6 Verification of Policy Generalisation

Although at any specific time a DRL agent with actor network μ—like the classifier—also returns an action $\mu(\mathbf{x})$ according to the state $\mathbf{x}$, the correctness of the action $\mu(\mathbf{x})$ is not solely dependent on the state $\mathbf{x}$. Instead, considering its training mechanism, the correctness of the action at any specific time depends on the expected long-term accumulated rewards. For this reason, the verification of a DRL agent also needs to consider not only the current state but also the long-term rewards (and therefore the future states). Therefore, to understand if a learned policy π works well in an environment MDP $\mathcal{M}^E$, we can conduct probabilistic model checking on their induced DTMC $\mathcal{M}^E(\pi, \mathbf{x}_0)$, where $\mathbf{x}_0$ is an initial state of $\mathcal{M}^E$. This enables the analysis of various properties that can be expressed with PCTL.

When the exact construction of the DTMC $\mathcal{M}^E(\pi, \mathbf{x}_0)$ is hard (due to e.g., some components of the environment $\mathcal{M}^E$ is unknown, or the policy network is too big for analysis), as discussed in [32], we can approximate it from a set of sampled trajectories.

Actually, due to the high-dimensionality of the underlying control problem and the continuity of some state and observable variables, it is unlikely that we can construct a DTMC that is exactly the application of policy π to MDP $\mathcal{M}^E$. Certain abstraction techniques [25] will be needed, for example, we consider predicate abstraction, where the DTMC's state space is constructed from a given set of predicates over MDP's state variables, as we will discuss below.

13.6.1 Construction of a DTMC Describing the Failure Process

We consider the execution of the policy π in an environment. For simplicity, we only differentiate the environments with a disturbance level that the robot's sensory input may be subject to, and assume that the disturbance level follows a distribution $\mathcal{N}(0, \sigma)$. Now, as stated in Sect. 13.1, given an MDP $\mathcal{M}^\sigma$ (based on a disturbance $\mathcal{N}(0, \sigma)$) and a DRL policy π, there is a trajectory distribution $\rho^{\pi,\sigma}(\zeta)$. Based on the *dynamics of risk-levels* in $\rho^{\pi,\sigma}(\zeta)$, we can define a DTMC, as shown in Fig. 13.2. It consists of a "negligible-risk" state s_N, a catastrophic failure state s_C, and several states s_{B_i} representing different levels of "benign failures".

Each trajectory is a sequence of successive states from the initial state to the end state of a DRL episode. First, we map each state in the trajectories to one of the states describing the failure process (i.e., s_N, s_C, and s_{B_i}). Second, we may conduct statistical analysis on the frequency of transitions between s_N, s_C, and s_{B_i}, based on which we estimate their corresponding transition probabilities. Finally, we construct the failure process DTMC with the defined structure and the estimated transition probabilities. To be exact, we describe the 3 main steps above as what follows.

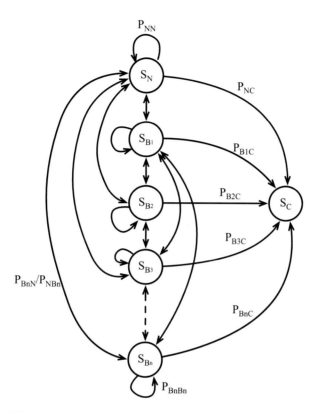

Fig. 13.2 The failure process DTMC based on risk-levels

13.6.1.1   Mapping MDP States onto DTMC States

First of all, every state in the DTMC (cf. Fig. 13.2) is associated with a risk level. Specifically, s_N is the negligible-risk state, s_C is the catastrophic failure state, and s_{B_i} are benign failure states such that the risk on s_{B_i} is higher than on s_{B_j} if $i > j$.

Now, to map S (the states on the trajectories) onto S (the states on the DTMC), we define a measure of risk based on the distance of the robot to obstacles. For instance, s_N suggests that the robot is 3+ m away from the obstacle, s_{B_1} suggests 2–3 m away, s_{B_2} suggests 1–2 m away, etc. Moreover, catastrophic failure s_C is defined as the robot terminated unexpectedly by a non-recoverable failure. The determination of the risk levels for states in S can be done by evaluating the sensory input. We remark that, this is a predicate abstraction where s_N, s_{B_1}, s_{B_2}, and s_C can be seem as predicates such that only one predicate can be True on any MDP state.

Definition 13.4 (Negligible-Risk Route) Given an MDP $\mathcal{M}^\sigma$ and a DRL policy π, a *negligible-risk route* is defined as a mission trajectory in $\rho^{\pi,\sigma}(\zeta)$ that contains only s_N states.

We remark that, the negligible-risk route is not necessarily the optimal route achieving the highest reward, rather it only depends on the risk-levels during the RAS mission.

13.6.1.2 Estimating Transition Probabilities

We can collect a set of mission trajectories by conducting statistical testing (the simple Monte Carlo sampling in our case) on $\rho^{\pi,\sigma}(\zeta)$. Then, all mission trajectories collectively can be transformed into a set of transitions, based on which we build a transition matrix to record the statistical data as follows:

	s_N	s_{B_1}	$\cdots$	s_{B_m}	s_C
s_N	$n_{1,1}$	$n_{1,2}$	$\cdots$	$n_{1,m+1}$	$n_{1,m+2}$
s_{B_1}	$n_{2,1}$	$n_{2,2}$	$\cdots$	$\cdots$	$\cdots$
$\cdots$	$\cdots$	$\cdots$	$\cdots$	$\cdots$	$\cdots$
s_{B_m}	$n_{m+1,1}$	$\cdots$	$\cdots$	$n_{m+1,m+1}$	$\cdots$
s_C	$n_{m+2,1}$	$\cdots$	$\cdots$	$\cdots$	$n_{m+2,m+2}$

where $n_{1,1}$ records the number of transitions from s_N to s_N, and so on. m is the number of levels of benign failures (that varies case by case depends on the application-specific context, e.g., we choose $m = 3$ in our experiments).

Let the transition probability matrix of the failure process DTMC be $\mathbf{P}_1 = (p_{ij}) \in [0, 1]^{(m+2)\times(m+2)}$. In a DTMC, given a current state i, the transition to a next state follows a *categorical distribution*. Due to the Markov property, the categorical distributions of each state are *independent*. Hence, as we observe repeated outgoing transitions from state i, the repeated categorical process follows a *multinomial distribution*. For the i-th row of $\mathbf{P}_1$, the likelihood function $\mathcal{L}$ is (by omitting the combinatorial factor):

$$\mathcal{L}(p_{i,1}, \ldots, p_{i,m+2} \mid n_{1,1}, \ldots, n_{1,m+2}) = \prod_{j=1}^{m+2} p_{i,j}^{n_{i,j}} \tag{13.11}$$

Upon establishing the likelihood function, many existing estimators can be invoked for our purpose, such as the basic Maximum Likelihood Estimation (MLE) and Bayesian estimators [47]. While more advanced estimators can be easily integrated in our proposed framework, we only present the use of MLE in this paper for brevity:

$$\hat{p}_{i,j} = \frac{n_{i,j}}{\sum_{j=1}^{m+2} n_{i,j}} \tag{13.12}$$

It is known that MLE is an unbiased estimator [47], while the uncertainty in the estimates is captured by the variance that depends on the number of samples. There are also means for calculating $(1 - \alpha)$ confidence intervals of the verification results, given the observations on the frequencies between states (exactly as our statistical data $n_{i,j}$). Such result may in turn determine the required number of samples $n_{i,j}$ given a required say 95% confidence level for the final verification results. Although we did not calculate the confidence interval to determine the sample size in this paper, we instead choose a sample size in our later experiments that is sufficiently large to show a converging trend of the verification results.

13.6.1.3   Construction of Failure Process DTMC

The failure process DTMC is the product of two DTMCs, M_1 and M_2, via the synchronisation of the transition actions.

Let $AP_1 = \{crash, neg\_risk, risk\_B_1, \cdots, risk\_B_n\}$, and we construct the first DTMC $M_1 = (S, s_N, \mathbf{P}_1, L_1)$ where

- $S = \{s_N, s_{B_1}, \ldots, s_{B_n}, s_C\}$,
- Each entry $p_{i,j}$ of $\mathbf{P}_1$ is defined as Eq. (13.12), and
- $neg\_risk \in L_1(s_N), risk\_B_i \in L_1(s_{B_i})$ for $i \in \{1..n\}$, and $crash \in L_1(s_C)$.

We also define a reward structure "deviation"$= (r_S, r_T)$ with

- $r_S(s_N) = 0$,
- $r_S(s_C) = 0$,
- $r_S(s_{B_i}) = d_i$ (where d_i is the deviation from s_N to s_{B_i}), and
- $r_T(s1, s2) = 0$ for all $s1, s2 \in S$.

Moreover, we need a "mission stage DTMC" (for simplicity, we only consider two stages—mission terminated or not). Let $AP = \{progressing, terminated\}$, we construct $M_2 = (K, k_0, \mathbf{P}_2, L_2)$ with

- $K = \{k_0, k_1\}$,
- $progressing \in L_2(k_0)$ and $terminated \in L_2(k_1)$, and
- The transition probabilities $\mathbf{P}_2$ are $p_{k_0,k_1} = \frac{1}{l_{mis}}$, $p_{k_0,k_0} = 1 - \frac{1}{l_{mis}}$, $p_{k_1,k_1} = 1$ and $p_{k_1,k_0} = 0$, where l_{mis} is a constant representing the expected mission length (number of transitions) obtained from the testing data.

We also define a reward structure for this DTMC: "step"$= (r_S, r_T)$ with

- $r_T(k_0, k_0) = 1$,
- $r_T(k_0, k_1) = 1$, and
- $r_S(k) = 0$ for all $k \in K$.

Finally, we encode the failure process DTMC with PRISM model checker [89].

13.7   Verification of State-Based Policy Robustness

The verification in Sect. 13.6 mainly concerns whether a learned policy works well in an environment. It actually computes the generalisation ability of the policy in the environment, considering the fact that the policy might not be trained on all paths of the DTMC but is expected to generalise well. Nevertheless, there are other safety properties to be considered. In this section, we consider temporal properties over the state-based robustness.

For state-based robustness, we concern the robustness of the policy network (and other supplementary network such as the critic network of DDPG) on individual states. Therefore, temporal properties over the state-based robustness measure the dynamics of such robustness of the neural networks when DRL agent interacts with the environment. Actually, state-based robustness is mainly due to the existence of observation noises (from e.g., noisy sensor reading), and we assume that such noise only influences the current state. We will discuss in the next section (Sect. 13.8) the case where the noise at a state may influence the future states.

Formally, to work with state-based policy robustness, we need to lift the model $\mathcal{M}^E(\pi, \mathbf{x}_0)$ into $\mathcal{M}^E(\pi, \mathbf{x}_0, C)$ to consider the constraint C on the noise.

13.7.1   Modelling Noise

Usually, the constraint $C(\mathbf{x})$ expresses the neighborhood of a state $\mathbf{x}$. A neighborhood can be e.g., a distance norm based neighborhood as

$$\{\mathbf{x}' \mid ||\mathbf{x} - \mathbf{x}'||_p \le d\}, \tag{13.13}$$

a Zonotope expressed as a Minkowski sum of a set of planes, or any other topological shape. Recall that, we use $|| \cdot ||_p$ to express the L_p norm of a vector for $p \ge 0$.

An immediate impact of the state-based noise is, instead of having a deterministic action and a single approximated Q value (when given a critic network), there are a set of possible actions and a set of possible approximated Q values.

13.7.2   Example Properties

We may consider a few example properties in the following, to quantify the uncertainty incurred by the noise.

Example 13.2 We may write property

$$\mathcal{P}_{\le 0} \, \lozenge \, (\mathbf{a} \in \mathcal{A}_{valid}) \tag{13.14}$$

to express that it is almost sure that some action **a** is never activated, where $\mathcal{A}_{valid}$ is the set of actions that are possible at the current state, subject to the noise. Moreover, we may write

$$\mathcal{P}_{\leq 0} \lozenge (r > c) \tag{13.15}$$

to express that the predictive Q value r (output by the critic network) is always no greater than c, subject to the noise at the current state.

For the remaining of this section, we will discuss how to extend the DTMC $\mathcal{M}^E(\pi, \mathbf{x}_0)$ into $\mathcal{M}^E(\pi, \mathbf{x}_0, C)$ to work with the properties as in Example 13.2. Actually, we need to expand the set *Prop* of atomic propositions to include atomic propositions such as $\mathbf{a} \in \mathcal{A}_{valid}$ and $r > c$. For this, we need to change the state space from $\mathcal{S}$ to $\mathcal{S} \times \mathcal{P}(\mathcal{S})$, so that we can compute e.g., $\mathcal{A}_{valid}$, r_{min}, and r_{max} over expanded states, where $[r_{min}, r_{max}]$ is the reachable range of the predictive Q value r.

13.7.3   Expansion of State Space

Comparing with $\mathcal{M}^E(\pi, \mathbf{x}_0)$, the new model $\mathcal{M}^E(\pi, \mathbf{x}_0, C)$ has an expanded state space $\mathcal{S} \times \mathcal{P}(\mathcal{S})$ such that each state $\mathbf{x}$ is now attached with a noise neighborhood $C(\mathbf{x})$. The transition relation depends only on the first element, i.e., on the state $\mathbf{x}$, so it can be easily obtained from the transition relation of the model $\mathcal{M}^E(\pi, \mathbf{x}_0)$. Another expansion is on the atomic propositions, which as explained will include some additional atomic propositions as in Example 13.2.

13.7.4   Evaluation of Atomic Propositions

Considering a typical DRL agent which has two neural networks $\mu : \mathcal{S} \to \mathcal{A}$ and $\nu : \mathcal{S} \times \mathcal{A} \to \mathcal{N}$, representing the actor and the critic agents, respectively. Intuitively, $\mu(\mathbf{x})$ returns the action **a** that needs to be taken on a state $\mathbf{x}$, while $\nu(\mathbf{x}, \mathbf{a})$ returns the predictive Q value of taking action **a** on state $\mathbf{x}$. Assume that we have a verification tool g that, given μ and a constraint $C(\mathbf{x})$ in $\mathcal{S}$, outputs an (over-approximated) reachable set in $\mathcal{A}$. That is, we can have

$$g(\mu, C(\mathbf{x})) \tag{13.16}$$

as the reachable set of actions when given a set of states expressed with constraint $C(\mathbf{x})$. With this, we can determine $\mathbf{a} \in \mathcal{A}_{valid}$ by knowing whether $\mathbf{a} \in g(\mu, C(\mathbf{x}))$.

Such verification tool is available [68] in previous works [144, 181, 98]. Moreover, we can compute

$$g(v, C(\mathbf{x}) \times g(\mu, C(\mathbf{x}))) \tag{13.17}$$

as the range of predictive Q value, and hence know the values r_{min} and r_{max}. Based on them, we can determine if $r > c$.

13.7.5 Probabilistic Model Checking Lifted with New Atomic Propositions

As explained above, when using neighborhood $C(\mathbf{x})$ to express adding noise to the state $\mathbf{x}$, we can lift the DTMC $\mathcal{M}^E(\pi, \mathbf{x}_0)$ into $\mathcal{M}^E(\pi, \mathbf{x}_0, C)$ by associating each state $\mathbf{x}$ with a neighborhood $C(\mathbf{x})$. Then, the probabilistic model checking proceeds the same as in Sect. 13.6 with the additional consideration of the neighborhood $C(\mathbf{x})$ on every state as well as the above-mentioned evaluation of atomic propositions.

13.8 Verification of Temporal Policy Robustness

The state-based policy robustness, as discussed in Sect. 13.7, concerns the dynamics of the robustness of the policy network μ (and the critic network v) on individual states of a path. When the states $\mathbf{x}_k$ and $\mathbf{x}_{k+1}$ are known, the robustness on state $\mathbf{x}_k$ cannot influence the robustness on $\mathbf{x}_{k+1}$. This is a simplistic assumption, but can be arguably unrealistic. In this section, we consider a more complex setting about the temporal evolution of robustness, i.e., the robustness on state $\mathbf{x}_k$ may influence the robustness of later states $\mathbf{x}_{k+1}, \mathbf{x}_{k+2}, \ldots$.

To work with temporal evolution of robustness, we cannot re-use the DTMC $\mathcal{M}^E(\pi, \mathbf{x}_0)$. Instead, we need to construct a new Kripke structure [25] $\mathcal{M}^E(\pi, C(\mathbf{x}_0)) = (S, s_0, T, L)$ from the environment MDP $\mathcal{M}^E$, the DRL policy π, the initial state $\mathbf{x}_0$, and the noise constraint C. Also, we do not work with the probabilistic logic PCTL because the transition relation T in M is non-probabilistic, and will work with a non-probabilistic temporal logic LTL.

13.8.1 State Space

Given a set of MDP states $\mathbf{X}$, we write $\eta(\mathbf{X})$ for the minimum polytope that contains the set $\mathbf{X}$, i.e., $\mathbf{X} \subseteq \eta(\mathbf{X})$, where η can be e.g., **Box** and **Zonotope**, as discussed above. Intuitively, $\eta(\mathbf{X})$ is an over-approximation of $\mathbf{X}$. Let S of $\mathcal{M}^E(\pi, C(\mathbf{x}_0))$ be the set of states such that every state s is a polytope $\eta(\mathbf{X})$. The function η is used to keep all the states in $\mathcal{M}^E(\pi, C(\mathbf{x}_0))$ as the same polytope shape, to make the state space manageable. With this definition of states S, the initial state $s_0 = \eta(C(\mathbf{x}_0))$ is the polytope containing the noise neighborhood of the initial observation $\mathbf{x}_0$.

13.8.2   Transition Relation

First of all, we need two supplementary functions. The first function h_1 is, given a constraint C, to split C into a set of constraints, i.e.,

$$h_1(C, g, \mu) = \{C^{\mathbf{a}_1}, \ldots, C^{\mathbf{a}_m}\} \tag{13.18}$$

such that $C = C^{\mathbf{a}_1} \cup \ldots \cup C^{\mathbf{a}_m}$ and, for all $1 \leq i \leq m$, $g(\mu, C^{\mathbf{a}_i}) = \{\mathbf{a}_i\}$. Intuitively, after splitting, all states in a constraint $C^{\mathbf{a}_i}$ take the same action $\mathbf{a}_i$. We write $a(C^{\mathbf{a}_i}) = \mathbf{a}_i$. The second function h_2 is, given constraints C and an action $\mathbf{a}$, to return the set of next states, i.e.,

$$h_2(C, \mathbf{a}) = \bigcup_{\mathbf{x} \in C} support(\mathcal{P}(\mathbf{x}, \mathbf{a})) \tag{13.19}$$

where $\mathbf{x} \in C$ means that the state $\mathbf{x}$ satisfies the constraint C, and $support(D)$ returns the support of a distribution D. We recall that $\mathcal{P}(\mathbf{x}, \mathbf{a})$ is the probabilistic transition function of $\mathcal{M}^E$. We remark that, h_1 can be obtained by applying binary search together with the verification tool g, and h_2 can be obtained by a genetic algorithm to search for a convex hull over the next states.

Based on h_1 and h_2, we can construct the transition relation T of $\mathcal{M}^E(\pi, C(\mathbf{x}_0))$. For every suitable polytope C, we let

$$(C, C') \in T \text{ for every } C' \in \{\eta(h_2(C^{\mathbf{a}}, \mathbf{a})) \mid C^{\mathbf{a}} \in h_1(C, g, \mu)\}. \tag{13.20}$$

Note that, with this construction, we assume that the noise is only from the initial state, without having any additional noise in the later steps. If we need to consider additional noise for every step, we may replace the set $\{\eta(h_2(C^{\mathbf{a}}, \mathbf{a})) \mid C^{\mathbf{a}} \in h_1(C, g, \mu)\}$ with $\{\eta(C(h_2(C^{\mathbf{a}}, \mathbf{a}))) \mid C^{\mathbf{a}} \in h_1(C, g, \mu)\}$, i.e., every input subset $h_2(C^{\mathbf{a}}, \mathbf{a})$ is added with noise represented with the constraint C.

13.8.3   LTL Model Checking

The labelling function L of $\mathcal{M}^E(\pi, C(\mathbf{x}_0))$ is the same as the one in the DTMC $\mathcal{M}^E(\pi, \mathbf{x}_0)$ for atomic propositions such as $\mathbf{a} \in \mathcal{A}_{valid}$ and $r > c$. The Kripke structure $\mathcal{M}^E(\pi, C(\mathbf{x}_0))$ is non-probabilistic. Therefore, we can apply LTL model checking [25] to understand the qualitative properties such as

$$\neg \Diamond \neg \, (\mathbf{a} \notin \mathcal{A}_{valid}) \tag{13.21}$$

which expresses that action **a** will never occur, and

$$\neg \Diamond \neg \, (r \le c) \tag{13.22}$$

which expresses that the predictive Q value r is always no greater than c.

13.9   Addressing Sim-to-Real Challenge

Considering the lack of real-world training data for DRL agents, it is a common practice to train a DRL policy in a simulation environment and then consider its application in real-world settings by transferring the knowledge and adapting the policy. More and more high-fidelity simulation platforms such as AirSim [151], CARLA [36], and RotorS [49] have been made available, in order to reduce the gap between simulation and reality. Nevertheless, the gap exists and technical means to detect and reduce the gap are still needed. Typical methods to reduce the gap during training include e.g., adding perturbances to the environment [197, 196, 173], building a precise mathematical model for a physical system [85, 81], domain randomisation [116], domain adaptation [5], policy distillation [170], and consideration of novel experiences that were not in the simulations [138].

Unlike the methods to reduce the gap, less works have been done on detection, i.e., identifying the gap and evaluating the extent of the gap. The verification method in Sect. 13.6 can be used to quantitatively evaluate how well a DRL policy π executes in an environment E. Because E can be the real world or the simulation environment, we can construct two models $\mathcal{M}^{E_{simu}}(\pi, \mathbf{x}_0)$ and $\mathcal{M}^{E_{real}}(\pi, \mathbf{x}_0)$, and use their difference such as

$$\mathcal{P}_{=?}^{E_{simu}}(\psi) - \mathcal{P}_{=?}^{E_{real}}(\psi), \tag{13.23}$$

to measure the gap between two environments E_{simu} and E_{real} over the probability of satisfying the property ψ. With this, the gap is exhibited roughly in the model view (E_{simu} vs E_{real}).

On the other hand, the verification methods in Sects. 13.7 and 13.8 concern the extent to which the environment noise, which may not appear in the simulation environment but will appear in the real world, may affect the properties. The sim-to-real gap can be exhibited by checking the difference between the probabilities of the properties before and after the consideration of environment noise. Similarly, this is also on model view (E_{simu} vs $E_{simu+noise}$).

However, none of the above methods is able to directly work with paths, to identify specific paths on which a path property ψ has different satisfiability results in different environments, or to compute the probability of such paths. We require that on such paths, the DRL agents take the same sequence of actions in different environment. Identifying these paths are essential, because they are direct evidence of the existence of gap between environments.

13.9.1   Model Construction

To enable our working with the above-mentioned paths, we can construct another DTMC $\mathcal{M}^{E_1 \times E_2}(\pi, (\mathbf{x}_0, \mathbf{x}_0)) = (S_1 \times S_2, (\mathbf{x}_0, \mathbf{x}_0), \mathbf{P}, L)$, which is a synchronisation of the two DTMCs $\mathcal{M}^{E_1}(\pi, \mathbf{x}_0) = (S_1, \mathbf{x}_0, \mathbf{P}_1, L_1)$ and $\mathcal{M}^{E_2}(\pi, \mathbf{x}_0) = (S_2, \mathbf{x}_0, \mathbf{P}_2, L_2)$ such that

- the transition relation is $P((\mathbf{x}_1, \mathbf{x}_2), (\mathbf{x}_1', \mathbf{x}_2')) = P_1(\mathbf{x}_1, \mathbf{x}_1') * P_2(\mathbf{x}_2, \mathbf{x}_2')$, with the requirement that $\mathcal{P}_1(\mathbf{x}_1' \mid \mathbf{x}_1, \mathbf{a}) > 0$ and $\mathcal{P}_2(\mathbf{x}_2' \mid \mathbf{x}_2, \mathbf{a}) > 0$ for some action $\mathbf{a}$. $\mathcal{P}_1$ and $\mathcal{P}_2$ are the probability transition relations of the MDPs $\mathcal{M}^{E_1}$ and $\mathcal{M}^{E_2}$, respectively. We note that, a normalisation may be needed to retain a probability distribution for every state $(\mathbf{x}_1, \mathbf{x}_2)$.
- the labelling function L will return a Boolean value for every atomic proposition on every state $(\mathbf{x}_1, \mathbf{x}_2)$, indicating whether or not the atomic proposition is evaluated the same on the two states $\mathbf{x}_1$ and $\mathbf{x}_2$, respectively, over their DTMCs.

13.9.2   Properties

Based on the construction of $\mathcal{M}^{E_1 \times E_2}(\pi, (\mathbf{x}_0, \mathbf{x}_0))$, we may verify some temporal properties such as

$$\neg \Diamond \neg (r > c) \tag{13.24}$$

which expresses that the robot always has the same evaluation on $r > c$ in two environments. Any counterexample to this will lead to a finite path where $r > c$ maintains the same evaluation until reaching a state $(\mathbf{x}_1, \mathbf{x}_2)$ where it is evaluated differently on $\mathbf{x}_1$ and $\mathbf{x}_2$, respectively. Probabilistic properties may also be considered, such as

$$\mathcal{P}_{=?} \neg \Diamond \neg (r > c) \tag{13.25}$$

which returns the probability of maintaining $r > c$ across the environments.

Chapter 14
Testing Techniques

Verification techniques, as discussed in Chaps. 10 and 13, are to ascertain—with mathematical proof—whether a property holds on a mathematical model. The soundness and completeness required by the mathematical proof result in the scalability problem that verification algorithms can only work with either small models (e.g., the MILP-based method as in Sect. 11.2) or limited number of input dimensions (e.g., the reachability analysis as in Sect. 11.3). In practice, when working with real-world systems where the machine learning models are large in nature, other techniques have to be considered for the certification purpose. Similar to traditional software testing against software verification, neural network testing provides a certification methodology with a balance between completeness and efficiency. In established industries, e.g., avionics and automotive, the needs for software testing have been settled in various standards such as DO-178C and MISRA. However, due to the lack of logical structures and system specification, it is less straightforward on how to extend such standards to work with systems with neural network components. In the following, we discuss some existing neural network testing techniques. The readers are referred to the survey [68] for more discussion.

14.1 A General Testing Framework

Assume that according to the model f and a safety property ϕ, we are able to define a set of test objectives $\mathcal{R}$. We will discuss later in Sect. 14.2 a few existing covering methods cov to define $\mathcal{R}$ for convolutional neural networks.

Definition 14.1 (Test Suite) Given a neural network f, a test suite $\mathcal{T}$ is a finite set of input instances, i.e., $\mathcal{T} \subseteq \mathcal{D}$. Each instance is called a test case.

X. Huang et al., *Machine Learning Safety*, Artificial Intelligence: Foundations, Theory, and Algorithms, https://doi.org/10.1007/978-981-19-6814-3_14

Ideally, given the set of test objectives $\mathcal{R}$ with respect to some covering method cov, we run a test case generation algorithm (to be introduced in Sect. 14.3) to find a test suite $\mathcal{T}$ such that

$$\forall \alpha \in \mathcal{R} \exists \mathbf{x} \in \mathcal{T} : cov(\alpha, \mathbf{x}) \tag{14.1}$$

where $cov(\alpha, \mathbf{x})$ intuitively means that the test objective α is satisfied under the test case $\mathbf{x}$. Intuitively, Eq. (14.1) means that every test objective is covered by some of the test cases. In practice, we might want to compute the degree to which the test objectives are satisfied by a test suite $\mathcal{T}$.

Definition 14.2 (Test Criterion) Given a neural network f, a covering method cov, a set $\mathcal{R}$ of test objectives, and a test suite $\mathcal{T}$, the test criterion $M_{cov}(\mathcal{R}, \mathcal{T})$ is as follows:

$$M_{cov}(\mathcal{R}, \mathcal{T}) = \frac{|\{\alpha \in \mathcal{R} | \exists \mathbf{x} \in \mathcal{T} : cov(\alpha, \mathbf{x})\}|}{|\mathcal{R}|} \tag{14.2}$$

Intuitively, it computes the percentage of the test objectives that are covered by test cases in $\mathcal{T}$ w.r.t. the covering method cov.

14.2 Coverage Metrics for Neural Networks

Research in software engineering has resulted in a broad range of approaches to testing software. Please refer to [203, 74, 157] for comprehensive reviews. In white-box testing, the structure of a program is exploited to (perhaps automatically) generate test cases. Structural coverage criteria (or metrics) define a set of test objectives to be covered, guiding the generation of test cases and evaluating the completeness of a test suite. E.g., a test suite with 100% statement coverage exercises all statements of the program at least once. While it is arguable whether this ensures functional correctness, high coverage is able to increase users' confidence (or trust) in the testing results [203]. Structural coverage analysis and testing are also used as a means of assessment in a number of safety-critical scenarios, and criteria such as statement and modified condition/decision coverage (MC/DC) are applicable measures with respect to different criticality levels. MC/DC was developed by NASA[59] and has been widely adopted. It is used in avionics software development guidance to ensure adequate testing of applications with the highest criticality [143].

Let $\mathcal{R}$ be the set of test objectives to be covered. For different structure coverage, we can define different sets of test objectives. In [131], $\mathcal{R}$ is instantiated as the set of statuses of hidden neurons. That is, for the set $\mathcal{H}_i$ of hidden neurons with ReLU activation functions at layer i, we let

$$\mathcal{R}_{\text{neuron coverage}} = \bigcup_{i=2}^{K-1} \mathcal{H}_i \tag{14.3}$$

Intuitively, combining with the general testing framework in Sect. 14.1, it requires to find a test suite $\mathcal{T}$ such that for every hidden ReLU neuron, there is a test case $t \in \mathcal{T}$ who can activate it. As another example, in [159], $\mathcal{R}$ is instantiated as the set of causal relationships between feature pairs. That is,

$$\mathcal{R}_{\text{MC/DC coverage}} = \bigcup_{i=2}^{K-2} \{(h_1, h_2) \mid h_1 \in \mathcal{H}_i, h_2 \in \mathcal{H}_{i+1}\} \tag{14.4}$$

Intuitively, combining with the general testing framework in Sect. 14.1, it requires to find a number of test pairs such that each $(h_1, h_2) \in \mathcal{R}_{\text{MC/DC coverage}}$, h_2 can be independently activated by h_1 [158].

14.3 Test Case Generation

Once a coverage metric is determined, it is needed to develop a test case generation algorithm to produce a set of test cases. Existing methods include e.g., input mutation [178], fuzzing [121], genetic algorithm [66], symbolic execution [160], and gradient ascent [161].

14.4 Discussion

In addition to testing techniques which originate from the software engineering area, there are other techniques that might be useful to analyse the safety of neural networks. Statistical evaluation applies statistical methods in order to gain insights into the verification problem we concern. In addition to the purpose of determining the existence of failures in the deep learning model, statistical evaluation assesses the satisfiability of a property in a probabilistic way, by e.g., aggregating sampling results. The aggregated evaluation result may have probabilistic guarantee, in the form of e.g., the probability of failure rate lower than a threshold l is greater than $1 - \epsilon$, for some small constant ϵ.

For the robustness, sampling methods, such as [177], are to summarise property-related statistics from the samples. While sampling methods can have probabilistic guarantees via e.g., Chebyshev's inequality, it is still under investigation on how to associate test coverage metrics with probabilistic guarantee. For the generalisation error, other than the empirical approach of using a set of test data to evaluate, recent efforts on complexity measure [21, 77] suggest that it is possible to estimate generalisation error—with theoretical bound—by only considering the weights of the deep learning without resorting to the test dataset.

Chapter 15
Reliability Assessment

In Chaps. 10, 13, and 14, we have introduced verification and testing techniques for convolutional neural network (CNN) and deep reinforcement learning (DRL). These techniques are to work with individual safety properties, such as the robustness of a data instance $\mathbf{x}$ (for CNN) or the safety of a policy π on a given initial state $\mathbf{x}_0$ (for DRL). However, considering that deep learning models usually serve as components of a large autonomous system, the safety of the deep learning models is related to how it is used in the system. For example, if a CNN is used as a perception component for e.g., object detection in a self-driving car, there will be a set D_{op} of data instances that may appear in operational time, all of which needs to be verified. If a DRL agent is used as a control component for e.g., navigation in a mobile robot, there will be a set of possible trajectories that may appear in operational time, all of which needs to be verified. In this chapter, we introduce a principled approach to utilise evidence produced by verification techniques about low-level safety properties (e.g., robustness of individual instances) to reason about high-level safety claims, such as "the perception component of the self-driving car can correctly classify the next 1000 instances with probability higher than 99%". These high-level safety claims are required in various industrial standards for software used in safety-critical systems. Methodologically, the approach is based on the safety argument, which provides a link between the safety evidence and a safety claim, showing that the safety evidence is sufficient to support the claim.

15.1 Reliability Assessment for Convolutional Neural Networks

Assume that, we are working with a verification technique g, which, given a network f and a constraint C, returns the probability of the inputs within C being classified correctly. The constraint C can be e.g., a norm ball with $\mathbf{x}$ as the centre to denote the possible perturbations.

Also, as mentioned above, we assume that there are a set D_{op} of operational data instances. We partition the input domain $\mathcal{D}$ into m cells, subject to the r-separation property [135]. These cells are disjoint and altogether form the entire input domain. Let p_{op} be the empirical distribution of the cells estimated with the dataset D_{op}. Then, we can learn a generative model G_{θ} over parameters θ such that

$$\theta^* = \arg\min_{\theta} \mathrm{KL}(G_{\theta}, p_{op}) \tag{15.1}$$

where $\mathrm{KL}(\cdot, \cdot)$ is the KL divergence between two distributions.

Based on the above, we can estimate the reliability (defined as the probability of failure in classifying the next input) as

$$\mathrm{Reliability}(f) = \sum_{i=1}^{m} G_{\theta}(C_i)(1 - g(f, C_i)) \tag{15.2}$$

Intuitively, $G_{\theta}(C_i)$ returns the probability density of the cell i represented as the constraint C_i, and $1 - g(f, C_i)$ returns the failure rate of the neural network f working on inputs satisfying the constraint C_i.

Zhao et al. [198, 200] show how to develop a principled safety argument to justify the reliability claim by aggregating evidence from either formal verification or statistical evaluation with Bayesian inference [156]. Other than learning a generative model, there are other methods to learn the distribution of cells [199].

15.2   Reliability Assessment for Deep Reinforcement Learning

The execution of a DRL-driven robot in an environment leads to a trajectory distribution (modelled as a distribute-time Markov chain, as discussed in Chap. 13), where the uncertainty (modelled with probability distribution) is from the environment.[1] Formally, given an environment E, a policy π, and an initial state $\mathbf{x}_0$, we can construct a model $\mathcal{M}^E(\pi, \mathbf{x}_0)$ representing the probability distribution of a set of trajectories. Assume that we have a verification technique g, as discussed in Sects. 13.6, 13.7, and 13.8.

Definition 15.1 The verification problem is, given a constructed model $\mathcal{M}^E(\pi, \mathbf{x}_0)$ and a verification tool g, to determine whether the model is safe with respect to

---

[1] For simplicity, we assume DRL policy is deterministic. There are DRL policies which are probabilistic, but the deterministic assumption is without loss of generality, and the methods we develop in this chapter can be easily adapted to work with probabilistic policies.

certain property ϕ, written as $\mathcal{M}^E(\pi, \mathbf{x}_0) \models^g \phi$. We may omit the superscript g and write $\mathcal{M}^E(\pi, \mathbf{x}_0) \models \phi$, if it is clear from the context. We can also assume that g returns a probability value—a Boolean answer can be converted into a Dirac probability. Then, the verification problem is to compute $Pr(\mathcal{M}^E(\pi, \mathbf{x}_0), \phi)$, i.e., the probability of safety.

In the following, we discuss how the above verification problem may contribute to the computation of the reliability. Similar as Sect. 15.1, we partition the set of possible initial states into m subsets, each of which is represented as a constraint C_i, for $i = 1..m$. Then, we can also define the empirical distribution p_{op} over the partitions, and find a model G_θ that is as close as possible to p_{op}. Formally, assume that G_θ is a generative model over parameters θ, we have

$$\theta^* = \arg\min_\theta \mathrm{KL}(G_\theta, p_{op}) \tag{15.3}$$

where $\mathrm{KL}(\cdot, \cdot)$ is the KL divergence between two distributions. Let $\mathbf{x}_{C_i}$ be the central point (i.e., a representative) of C_i.

Based on these, we can estimate the reliability (defined as the probability of failure in satisfying ϕ with the policy π in the environment E) as

$$\mathrm{Reliability}(E, \pi, \phi) = \sum_{i=1}^{m} G_\theta(C_i)(1 - Pr(\mathcal{M}^E(\pi, \mathbf{x}_{C_i}), \phi)) \tag{15.4}$$

where $G_\theta(C_i)$ returns the probability density of the partition i that is represented as the constraint C_i, and $1 - Pr(\mathcal{M}^E(\pi, \mathbf{x}_{C_i}), \phi)$ returns the failure rate of the DRL agent π working on inputs satisfying the constraint C_i under the environment E.

Note that, G_θ can be estimated in the same way as the data distribution in the convolutional neural networks. The discussion on the computation of $\mathcal{M}^E(\pi, \mathbf{x}_0) \models \phi$ or $Pr(\mathcal{M}^E(\pi, \mathbf{x}_0), \phi)$ is in Chap. 13.

Chapter 16
Assurance of Machine Learning Lifecycle

Up to now, all the techniques discussed are working with subjects at certain stage of the machine learning development cycle. For example, verification techniques are working with trained models, and adversarial training is working with models when learning. However, to ensure the safety of critical systems, safety assurance is usually required to assure the development lifecycle and demonstrate to others (such as third party clients and authorities) that the system performs accordingly. In general, safety assurance activities include systematic processes for continuous monitoring and recording of the system's safety performance, as well as evaluation of the safety processes and practices. In terms of the machine learning systems, safety assurance activities are to monitor, evaluate, and enforce safety measures for their lifecycle. Figure 16.1 presents the four stages in machine learning cycle when considering their working in safety critical systems: data preparation, model construction and model training, verification and validation, and runtime enforcement. In this chapter, we will discuss some perspectives (e.g., good practice, safety measurement) for each lifecycle stage, as indicated in Fig. 16.1. For example, for data preparation stage, we will present a workflow of good practice in preparing the training dataset, and discuss a measurement on determine its quality (i.e., the sufficiency). For runtime enforcement, we will discuss a few aspects that are essential to the safe deployment of the machine learning model in an environment, focusing on the partnership of AI and humans. We also discuss expected outcome of each stage.

16.1 Assurance on Data

Most efforts on the analysis of machine learning models, as we discussed in previous chapters, are on the trained models. This is based on an assumption that, the trained models have taken into consideration all necessary information from the training dataset; no more, no less. Any deviation from this assumption may lead to the

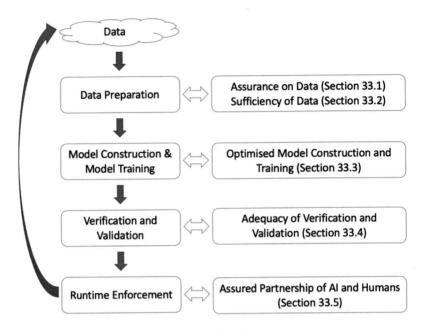

Fig. 16.1 Safety assurance of machine learning lifecycle

analysis results not applicable to the machine learning development cycle. For example, if the generated test cases are not on the same distribution with the training data, then the reliability assessment results (Sect. 15) such as the failure rate based on the test cases will not be valid. If the training dataset does not conform with the operational data distribution, then the empirical generalisation error will not be a good approximation to the true generalisation error (Sect. 3.1). There are also some safety vulnerabilities such as data poisoning that cannot be detected simply from a trained model. While an exact computation, or even an estimation, on the deviation is hard to achieve, we believe a few data processing steps are useful as *good practice* to improve the quality of training data and are therefore essential for safety assurance of machine learning. These steps include data cleaning, data pre-processing, data augmentation, and data anomaly detection. Figure 16.2 provides an illustrative diagram showing the flow of data in these steps.

Data cleaning is a process of ensuring that data is correct, consistent and usable. There are some recent good practices in industry. For example, [149] discusses a good practice in Amazon on automating the data validation process, where the quality of data is measured from three aspects: completeness, consistency, and accuracy. The completeness refers to the degree to which a data instance includes data required to describe a real world object, the consistency is defined as the degree to which a set of semantics rules are violated, and the accuracy is the correctness of the data. A set of pre-defined constraints are utilised by the user to define more

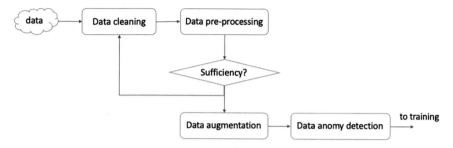

Fig. 16.2 Assurance on data

involved data quality constraints, over which one can determine the quality of a dataset.

After passing the data cleaning, the dataset is *pre-processed* with a few typical methods, such as standarisation, normalisation, whitening, and decorrelation. These techniques are to optimise the dataset in order to help machine learning algorithms to achieve better performance.

The data cleaning and pre-processing steps will result in a dataset that can be utilised for training. However, we still need to ensure that, for a target machine learning algorithm, the data is sufficient for training a good model. Different machine learning algorithms may have different data sufficiency requirements. A principled decision process (to be discussed further in Sect. 16.2) is required to determine the sufficiency of training data. If it is believed that the training data is insufficient, additional efforts may be needed on data collection. On the other hand, for some safety properties such as the robustness, additional data may come from *data augmentation*, which generates more data during the training phase so that the resulting trained model performs better on the safety property. However, unlike safety properties, extra care is needed to use data augmentation for the improvement of the accuracy of the model, because it will risk losing the i.i.d. property of training data.

Finally, related to the safety of machine learning, we need to be mindful on the potential of data poisoning and backdoor attacks, and it is recommended that some techniques (as we discussed in the previous chapters) for the detection and reduction of such risks would be desirable. Along this line, [15] considers a good practice at Google on the quality of data that will be fed into machine learning pipeline. It mainly focuses on utilising an anomaly detector to deal with challenges such as unexpected patterns, schema-free data, and training/serving skew.

Expected Outcome

The resulting dataset is required to be complete, consistent, accurate, optimised, sufficient, and free from outliers and poisoning data. All these properties can, and should, be objectively evaluated. We remark that, there are other requirements, such as balanced data and data diversity, which are also desirable to have for the safety assurance. If additional requirements are to be imposed, objective measurements and validation methods are needed.

16.2   Sufficiency of Training Data

Given a target machine learning algorithm and a set of training data, it is imperative to understand if the training dataset is sufficient for the training of a good machine learning model. While there are some rules of thumb that may be applicable, for example, the training data needs to be 10 times over the number of trainable parameters, it is believed that more principled methods are needed. We remark that, when the training data is insufficient for a given machine learning algorithm, it is easy to make the resulting trained model overfitted, although there are some recent discussions on the observation that over-parameterised neural network does not overfit [1].

VC-Dimension

First of all, theoretical results such as Vapnik–Chervonenkis dimension (or VC-dimension) are available to roughly estimate the required number of training data. Specifically, VC-dimension measures the capacity (or expressive power) of a machine learning model. It is defined as the cardinality of the largest set of data points that the machine learning algorithm can shatter. Formally, a machine learning model f parameterised with θ is said to shatter a set of data instances if for all assignments of the labels to the data instances, there exists a valuation of θ such that f makes no error when classifying the set of data instances. The readers are referred to Sect. 10.1.3 for an intuitive explanation on the expressivity of a perceptron. For a perceptron of k variables (for example, $k = 2$ as in Table 10.2), the VC-dimension is $k + 1$.

Once we compute the VC dimension d_{VC} for a model f, we have the following VC-bound [171]:

$$Pr\left(Err(f, \mathcal{D}) \leq Err(f, D_{train}) + \sqrt{\frac{1}{N}[d_{VC}(\log(\frac{2N}{d_{VC}})+1) - \log(\frac{\epsilon}{4})]}\right) = 1 - \epsilon$$

(16.1)

where $1 - \epsilon$ is an arbitrary probability expressing the confidence, and N is the number of data instances. Then, according to Eq. (16.1), given the required error

$Err(f, \mathcal{D})$ and confidence $1 - \epsilon$ from the problem, we are able to estimate the number N of training data instances.

We remark that, in addition to VC-dimensions, there are other machine learning theoretical methods, such as Rademacher complexity and PAC Bayes theory, that can be utilised for this purpose.

Learning Curve

In addition to the estimation through VC-dimension, there are empirical ways to determine the size of training dataset. For example, learning curves (as we explained in Fig. 2.1 as an example) presents the model performance as the function of the training dataset size. By plotting the learning curve, we are able to monitor the convergence of the curve. Once the learning curve is converged (with e.g., an ϵ termination threshold), the increase of training dataset size will not lead to significantly improved performance, and it is regarded as the sufficiency of the training data.

16.3 Optimised Model Construction and Training

In this section, we focus on the feedforward neural networks. Once the training dataset is ready, the developer needs to construct a neural network, decide on the training hyper-parameters, and then apply optimisation algorithm to train the neural network. In the following, we start with an analytical analysis of them through a decomposition of the generalisation error (Sect. 3.1).

We write $G_f^{0-1} = GE(f, D_{train}, D_{test})$ for the 0-1 generalisation error, and N for the set of possible neural networks. Then, G_f^{0-1} can be decomposed as follows:

$$G_f^{0-1} = \underbrace{G_f^{0-1} - \inf_{f \in N} G_f^{0-1}}_{\text{Estimation error of } f} + \underbrace{\inf_{f \in N} G_f^{0-1} - G_{D_{train}}^{0-1,*}}_{\text{Approximation error of } N} + \underbrace{G_{D_{train}}^{0-1,*}}_{\text{Bayes error}} \qquad (16.2)$$

where $G_{D_{train}}^{0-1,*}$ is the 0-1 generalisation error of the Optimal Bayes classifier. The *Bayes error* is the lowest and irreducible error over all possible classifiers for the given classification problem [48]. It is non-zero if the true labels are not deterministic (e.g., an image being labelled as y_1 by one person but as y_2 by others), thus intuitively it captures the uncertainties in the dataset D_{train} and the true distribution $\mathcal{D}$ when aiming to solve a real-world problem with machine learning. The *approximation error of* N measures how far the best classifier in N is from the overall optimal classifier, after isolating the Bayes error. The set N is determined by the architecture of the machine learning model, thus lifecycle activities at the **model**

construction stage are used to minimise this error. Finally, the *estimation error of f* measures how far the learned classifier *f* is from the best classifier in N. Lifecycle activities at the **model training** stage essentially aim to reduce this error.

Both the approximation and estimation errors are reducible. The *ultimate goal* of all lifecycle activities is to reduce the two errors to 0, especially for safety-critical systems. This is analogous to the "possible perfection" notion of traditional software as pointed to by Rushby and Littlewood [101, 147]. That is, assurance activities, e.g., performed in support of DO-178C, can be best understood as developing evidence of possible perfection. Similarly, for safety critical machine learning model, we believe its lifecycle activities should be considered as aiming to train a "possibly perfect" model in terms of the two *reducible* errors. Thus, we may have some confidence that the two errors are both 0 (equivalently, a prior confidence in the *irreducible* Bayes error since the other two are 0), which indeed is supported by on-going research into finding globally optimised DNNs [37].

16.3.1   Neural Architecture Search

While it is often regarded as dark-art for the model construction and training because they are usually dependent on the experience of the developer, it is not hard from the above discussion that they can be treated as an optimisation problem to reduce the approximation and estimation errors. Towards this, neural architecture search [204] has been recently discussed through e.g., reinforcement learning to automatically generate high-performing neural network architectures for a given learning task. The reinforcement learning agent explores a large but finite space of possible architectures and iteratively discovers designs with improved performance on the learning task.

16.3.2   Best Practice

Neural architecture search requires significant computational resources so may not be applicable to all developments. An alternative way is to follow the best practice in machine learning development, which includes the following steps:

1. Start with a simple model with limited complexity (e.g., small number of features and restricted model capacity). Repeatedly do the following:
   (a) tune the hyper-parameters to train a model, and
   (b) apply debugging tools (e.g., verification and validation tools for safety properties, and the evaluation methods as introduced in Chap. 2) to understand the performance,

   until finding a valid model. Take this model as a baseline model.

2. Repeatedly do the following:

    (a) add complexity (by e.g., increasing the number of features, and increasing model capacity) to a baseline model, and

    (b) Repeat step 1a and 1b until obtaining a new trained model that performs better than the baseline model. We need to make sure than a more complex model should always perform better than a less complex model. Make the new model as a baseline model.

Note that, the above step 2 may lead to multiple baselines, and we always need to justify the benefit of added complexity with improved performance.

3. Select a model from a set of baseline models according to the required, acceptable level of performance.

It is not hard to see that, the above best practice is also a controlled optimisation process to construct a more and more complex model by gradually increasing the model complexity. Different from the neural architecture search, this process has human in the loop, where the developer needs to design the added complexity and tune the hyper-parameters.

16.3.3 Expected Outcome

The outcome of the optimised model construction and training is a well trained model with minimum generalisation estimation and approximation error. The errors can be estimated in an empirical way with a test dataset or some more principled methods such as complexity measures [77]. In addition, as indicated in Part 12, there might be other requirements, including the safety properties we discussed in Sect. 3, that are needed to be considered in model construction and training. Objective measurements are needed to determine if they are properly implemented.

16.4 Adequacy of Verification and Validation

This section discusses when a verification and testing technique is able to conclude the sufficiency, or completeness, of the analysis. For example, when using testing method, we need to know when to terminate the test case generation process. When applying robustness verification, we need to know how many local instances we need to verify. We remark that, the discussion in this section is mainly concerned with the high-level safety assurance, considering the operational use of a machine learning model that may have a set of different inputs from a data distribution $\mathcal{D}_{op}$ (see similarly in Sect. 15.1). It is different from the robustness verification, in which as discussed in [68] and Part III the completeness is mainly concerned about the exhaustiveness of the input instances in a small neighbourhood around a given input instance.

There are mainly two approaches that can be utilised, including the behaviours of a machine learning model in inference stage and the data instances that might appear in operational stage, respectively. For these approaches, objective metrics are needed to determine the extent to which an analysis technique has conducted, as discussed in the later ALARP principle.

16.4.1   Coverage of Machine Learning Behaviours

While usually treated as "black-box", most machine learning models have internal behaviours when processing a data instance. It has been noted in [66] that, even for the complex recurrent neural networks, two input instances with the same internal behaviours, defined as the temporal evolutions of the joint latent representation of all gates and internal states, will represent the same instance and get the same classification result. For convolutional neural networks, the internal behaviours can be the vectors of latent representations of different layers. Therefore, the exploration of all internal behaviours will be adequate for the verification and testing. However, due to the continuous nature, the number of behaviours can be infinite, which suggests that some level of abstractions are usually needed to define behaviours.

The abstracted behaviours include low-level ones, such as the activation of individual ReLU neurons and the activation of causality relation between neurons, and high-level ones, such as the semantics relations between activation vectors of different layers. The low-level ones have led to the proposals of various structural coverage metrics, such as neuron coverage and MC/DC coverage as we discussed in Sect. 14.2, while the high-level ones have led to other proposals, such as the semantics abstraction of the neural networks as in Chap. 17.3 and the symbolic representation of the temporal evolution of the latent representations as in [66]. For the latter, once a semantics representation is defined, the metric is to measure the percentage of possible concrete semantics instances that have been explored for the verification and testing.

We remark that, for both low-level and high-level behaviours, their metrics might not be able to reach 100% coverage, because some behaviours can be infeasible for the machine learning model. Therefore, instead of setting up threshold for the termination of analysis, empirical methods, such as a similar technique as the learning curve as discussed in Sect. 16.2, will be needed to determine whether the analysis has been adequate with respect to the metric.

A more intriguing observation is that, these metrics might not be tight enough to study a machine learning model, for example, certain neuron activation, defined as a behaviour of a convolutional neural network processing images, may appear multiple times when a neural network works with different input instances. This is mainly due to the fact that the abstracted behaviour is too coarse. Therefore, the definition of behaviours need to be carefully designed so that it strikes a balance between adequacy and complexity.

16.4.2 Coverage of Operational Use

An alternative way of considering the adequacy of verification and testing is to explore the set of all possible input instances. That is, if we are able to enumerate all instances that may appear when the machine learning model is used, and confirm their safety, we are certain about the safety of the machine learning model.

However, due to the missing of the true data distribution $\mathcal{D}$, it is unlikely that we are able to directly enumerate the operational data instances. To deal with this, methods such as variational autoencoder and generative adversarial network can be utilised to learn the data distribution. Based on the learned data distribution, a set of seeds can be selected for the analysis, as discussed in [200].

Moreover, the direct working with data distribution $\mathcal{D}$ enables the possibility of integrating human prior knowledge. Intuitively, human experts may have a good level of knowledge that certain features can be more important (and therefore should be weighted higher) in an operational environment than another. For example, snowing on images can be more relevant for Toronto than California. Such prior knowledge can be integrated into the determination of the coverage of operational use (e.g., the learned data distribution) so that the resulting verification and validation is contextually relevant.

16.4.3 ALARP ("As Low As Reasonably Practicable")

For both the above coverage methods, the continuity and high-dimensionality of the spaces to be covered make the exhaustive enumeration unlikely. Therefore, certain adaptation of the ALARP ("As Low As Reasonably Practicable") principle might be helpful to strike a cost-benefit balance. Principled approaches are needed to weight the safety risks and measure the cost needed to identify the risks. Then, a monitoring process runs in parallel with the verification and validation process to determine when the cost involved in identifying the risks would be grossly disproportionate to the benefit gained.

16.4.4 Expected Outcome

The outcome of the adequacy of verification and validation is either a proof of the adequacy or a validation report containing objective measurement on the adequacy. A justification on the cost-benefit balance is also needed.

16.5   Assured Partnership of AI and Humans

Up to now, all the topics are based on data, algorithms, and models, with the interactions with humans mainly on requiring humans to e.g., label the training instances, and differentiate whether a perturbed image is an adversarial example. These interactions completely rely on humans' functional ability, without utilising their social aspects of intelligence. However, AI systems, or software/hardware systems with AI components, are penetrating our everyday life. It is therefore imperative to consider the partnership of humans and AI. From safety perspective, when humans and AI are interacting, the safety issues become more prominent, because unexpected behaviours of the AI systems may easily lead to bad consequences on their human partners.

In this section, we will discuss several technical topics on how to assure the partnership of AI with humans. They include requiring a machine learning agent to provide auxiliary information beyond decisions, such as the confidence of each decision (i.e., *uncertainty estimation*) and the explanation of individual decisions and the model in general (i.e., *explainability and interpretability*), and understand human context (i.e., *contextualisation*). These technical means can help strengthen the trust between humans and AI, and therefore enable their safe co-operation.

16.5.1   Uncertainty Estimation

Essentially, a machine learning model is a function that is obtained by minimising a given loss function over a set of training data. The function may behave wrongly, which can be evaluated with verification and validation techniques or detected with runtime monitoring techniques. In addition to the misbehaviour, another key concern is on the confidence of the machine learning model in making the wrong decision. It is clear that the situation is worse if a misbehaviour is conducted with high confidence than with low confidence. The confidence naturally comes with the Bayesian view on learning, where a deterministic trained model (such as a trained neural network) is seen as a sample from a distribution of models. Uncertainty estimation on machine learning models has been intensively discussed, with some existing methods such as MC dropout [50] and deep ensemble [91] that can be applied for neural networks.

The evaluation of uncertainty estimation methods is non-trivial, mainly due to the lack of "ground-truth" uncertainties. It is therefore useful to evaluate against a set of concrete baseline datasets and evaluation metrics that cover all types of uncertainties. In addition, a typical measurement regarding risk-averse and worst case scenarios is usually considered. Specifically, it requires that uncertainty predictions with a very high predicted uncertainty should never fail.

16.5.2 Explainability and Interpretability

To gain trust from human users in operational use, it is essential to enable them to understand the decisions a machine learning model has made. Explainable AI is a topic with fierce discussions lately. The reader is referred to recent surveys such as [68] for more explanations on the recent progress in this direction. There are a number of principles that have been frequently used to evaluate whether an explanation is good, such as accuracy, fidelity, consistency, stability, comprehensibility, certainty, and degree of importance [72].

16.5.3 Contextualisation

Contextual AI is a collection of techniques aiming to embed human context into AI systems so as to enable their interaction with humans. First of all, context-awareness requires that the AI systems are able to "see" at the same level as a human does. Then, based on the *human-level observation*, contextual AI is capable of analysing the cultural, historical, and situational aspects surrounding incoming data, and synthesising a context that makes the most sense to the humans. Such *human-level reasoning* enables the contextual AI to have the sufficient understanding about the human's environment, situation, and context. It is based on this level of understanding that it is able to explain, reason, behave, and collaborate with the human.

Unfortunately, neither the learning algorithms nor the analysis techniques naturally have human-level observation and reasoning. For example, for image classification task, both the deep learning algorithms and their analysis techniques are primarily based on pixels, while humans understand the images through high-level features. Typical ways of enhancing machine learning algorithms with contextualisation can be done through e.g., apply explainable AI techniques to obtain human-level observation, synthesise context into structures such as knowledge graph and Bayesian network, and then conduct human-level reasoning (e.g., logic reasoning, commonsense reasoning, or probabilistic inference) over the synthesised structures.

Analysis techniques, such as verification and validation techniques, should also been lifted to human-level observation and reasoning. Actually, the consideration of pixel-level analysis techniques has led to the notorious scalability issues that verification techniques are only able to work with either small-size neural networks (as in Sect. 11.2) or limited number of input dimensions (as in Sect. 11.3). Even for testing techniques, the scalability is also an issue when considering tighter coverage metrics such as MC/DC (Sect. 14.2). Therefore, the analysis techniques to support the reasoning on higher-level features are needed to not only make the analysis make sense to humans but also focus the limited analysis cost on the most important aspects.

Another perspective is on the risk of de-contextualisation in terms of the choice of models. Models originally used for one purpose may not be suitably re-used in a different context and for a different purpose. Validation activities are required to understand the impact of contextual changes on the safety.

16.5.4 Expected Outcome

The outcome of assured partnership is a machine learning model that is able to perceive, reason, and behave at the same level as its human partners. The three perspectives discussed above (uncertainty estimation, explainability and interpretability, and contextualisation) are essential for this purpose. In addition to their respective evaluations in particular for the uncertainty estimation and explainability and interpretability, a holistic evaluation on the assured partnership, or more formally trust between humans and AI, should also be considered. The trust evaluation is needed to be supported by rigorous reasoning frameworks such as [69], where the trust is quantitatively measured, and dynamically updated with the interactions, to enable verification techniques to be applied. Empirical experiments based on these theoretical frameworks should be conducted to validate the success of partnership.

Chapter 17
Probabilistic Graph Models for Feature Robustness

Up to now, we have known that machine learning algorithms can be used to effectively learn a function f from a set of input-output pairs. The function f approximates the relation between two random variables X and Y, and actually expresses the conditional probability $P_f(Y|X)$. However, in a complex, real world system, there might be more than two random variables and it is useful to not only understand the conditional probability between random variables but also be able to infer more intriguing information from the conditional dependence relations between random variables. It is also possible that, a complex machine learning model, such as a convolutional neural network, can be approximated by constructing the conditional dependencies between a set of random variables representing the features learned by the neural networks, see e.g., [10] for an example. Therefore, while it is agreeable that machine learning has been able to support human operators in dealing with some long-standing tasks such as object detection and recognition with accuracy and efficiency, there is still a need to infer useful knowledge from a set of conditional probabilities. This chapter is to explain how we may utilise probabilistic graphical model, a formalism to express conditional probabilities between random variables (X, Y, and variables for latent representations), as an abstract model for neural network. We will present the definition in Sect. 17.1 and then discuss the abstraction method in Sect. 17.3. More detailed introduction to probabilistic graphical models is given in the Appendix C.

17.1 Definition of Probabilistic Graphical Models

Probabilistic graphical models are a formalism for the above purpose. They use a structure, or more specifically a graph, to represent conditional dependence relations between random variables, and use a probability table for every random variable to express the local dependence relation of the random variable. There are two major

X. Huang et al., *Machine Learning Safety*, Artificial Intelligence: Foundations,
Theory, and Algorithms, https://doi.org/10.1007/978-981-19-6814-3_17

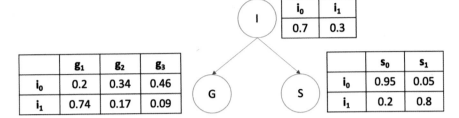

Fig. 17.1 A simple graphical model of three nodes

branches of graphical models, namely, Bayesian networks and Markov random fields, and in this chapter, we focus on Bayesian network. In the following, we will use (probabilistic) graphical models and Bayesian network interchangeably.

Depending on the dependence relation of individual random variable, the probability tables in a graphical model can be a marginal probability table, which shows that the random variable does not depend on any other variables in the graph, or a conditional probability table, which represents the conditional probability distribution of the current random variable over other random variables in the graph.

Figure 17.1 presents a simple graphical model with 3 nodes: I, G, S. We can see that one of the nodes I has a marginal probability table while the remaining two nodes have conditional probability tables. Summarising,

$$\text{Probabilistic Graphical Model} = \text{Graphical Structure} + \text{Multivariate Statistics} \tag{17.1}$$

Formally, a probabilistic graphical model $G = (\mathcal{V}, \mathcal{E}, P)$ where $\mathcal{V}$ is a set of nodes, representing the random variables, $\mathcal{E}$ is the set of edges between nodes, and P is a set of probability tables, one for each node in $\mathcal{V}$.

17.2 A Running Example

Assume that, on a self-driving car, there are two sensors, *Camera* and *Radar*, that are used to detect pedestrian collectively. The precision of the camera may be affected by weather conditions, such as the *Fog* as we consider in this example. The *Radar* may be affected by the distance of the object from the car, i.e., it can be very precise when the object is close but may become less precise when the object is *Away*. Once a pedestrian is detected and it is not away, the car will need to stop.

Figure 17.2 presents a probabilistic graphical model for this example. In the graph G, there are six random variables: *Camera, Radar, Fog, Detected, Away*, and *Stopped*. Every node is associated with either a marginal probability table or a conditional probability table, depending on whether they have incoming edges. The information about the probability tables are given in Fig. 17.3. For example,

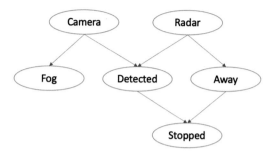

Fig. 17.2 A simple Bayesian network for safety analysis on vehicle stopping upon pedestrian detection

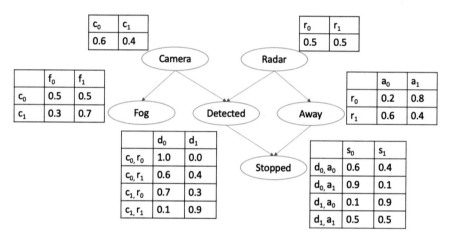

Fig. 17.3 Probabilistic table of the graphical model in Fig. 17.2

the nodes *Camera* and *Radar* do not have incoming edges, so each of them is associated with a marginal probability table. Intuitively, the two tables suggest that the probability of a pedestrian appearing in the imagery input of the camera is 0.4, and in the signal input of the radar is 0.5. Note that, the "appearing" is for ground truth (through human's eyes), not for the result of a detection system. The detection is implemented through the *Detected* node to be explained below.

Other nodes are associated with conditional probability tables. For example, the table for *Fog* shows that, when there is no pedestrian appearing in the imagery input, the probability of the foggy weather condition is 0.5. This probability is lowered when there is a pedestrian appearing in the imagery input. This is intuitive, because the foggy condition may affect the ability of camera capturing the pedestrian. Similar for the *Away* node. When there is no pedestrian appearing in the signal input, the probability of its away from a pedestrian is 0.8. This probability is lowered to 0.4 when there is a pedestrian appearing in the signal input.

The detection result is a fusion of both camera and radar's results. Note that, even if a pedestrian appears in the imagery input, it does not mean that the pedestrian can be detected (Recall the generalisation error and robustness error of deep learning). We note that, if neither of the sensors has a pedestrian appeared, no detection can be made at all. If one of the sensors has a pedestrian, there is a non-trivial chance that it can be detected. The detection becomes significantly better when both sensors captured the pedestrian.

Finally, the decision making on whether the car should be stopped is based on both the detection result and the distance. If it is detected (i.e., $Detected = d_1$) and not far away (i.e., $Away = a_0$) then this probability is high (0.9). Otherwise, the probability is low (0.1). The lowest probability appears when no pedestrian is detected (i.e., $Detected = d_0$) and it is away (i.e., $Away = a_1$).

Where Does Machine Learning Play a Role?

Machine learning can be used to generate those conditional probability tables. For example, a deep learning model can be designed and trained to get the table for *Detected* node, i.e., classify whether a pedestrian is detected or not on both the camera input and the radar input. Similarly, other nodes such as *Fog*, *Away*, and *Stopped* may also be implemented with a machine learning model.

17.3   Abstraction of Neural Network as Probabilistic Graphical Model

In this section, we construct a Bayesian network out of a trained neural network. In the end, the Bayesian network captures the distribution of neuron valuations in terms of latent features encoded in each neural network layers, as well as their causal relationships.

Extraction of Hidden Features

Assume that some feature extraction technique such as PCA and ICA has been used to analyse the neuron activation vector $\mathbf{v}_i$ of layer i that are induced by a given *training set* D_{train}. This produces a set of feature mappings $L_i = \{\lambda_{i,j}\}_{j \in \{1,...,t_i\}}$ for t_i features, such that each $(\lambda_{i,j} : L_i \to F_{i,j})$ maps the vector space L_i of neuron valuation into the j-th component of the feature space $F_{i,j}$.

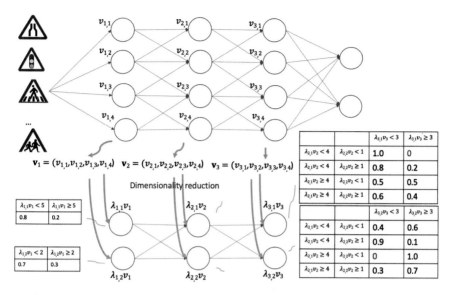

Fig. 17.4 Reducing neural networks to Bayesian networks

$\mathbf{v}_1 = (v_{1,1}, v_{1,2}, v_{1,3}, v_{1,4})$ $\mathbf{v}_2 = (v_{2,1}, v_{2,2}, v_{2,3}, v_{2,4})$ $\mathbf{v}_3 = (v_{3,1}, v_{3,2}, v_{3,3}, v_{3,4})$

Dimensionality reduction

$\lambda_{1,1}v_1 < 5$	$\lambda_{1,1}v_1 \geq 5$
0.8	0.2

$\lambda_{1,2}v_1 < 2$	$\lambda_{1,2}v_1 \geq 2$
0.7	0.3

		$\lambda_{3,1}v_3 < 3$	$\lambda_{3,1}v_3 \geq 3$
$\lambda_{2,1}v_2 < 4$	$\lambda_{2,2}v_2 < 1$	1.0	0
$\lambda_{2,1}v_2 < 4$	$\lambda_{2,2}v_2 \geq 1$	0.8	0.2
$\lambda_{2,1}v_2 \geq 4$	$\lambda_{2,2}v_2 < 1$	0.5	0.5
$\lambda_{2,1}v_2 \geq 4$	$\lambda_{2,2}v_2 \geq 1$	0.6	0.4

		$\lambda_{3,2}v_3 < 3$	$\lambda_{3,2}v_3 \geq 3$
$\lambda_{2,1}v_2 < 4$	$\lambda_{2,2}v_2 < 1$	0.4	0.6
$\lambda_{2,1}v_2 < 4$	$\lambda_{2,2}v_2 \geq 1$	0.9	0.1
$\lambda_{2,1}v_2 \geq 4$	$\lambda_{2,2}v_2 < 1$	0	1.0
$\lambda_{2,1}v_2 \geq 4$	$\lambda_{2,2}v_2 \geq 1$	0.3	0.7

Actually, L_i is such that the neuron values $\mathbf{v}_i$ for any input $\mathbf{x} \in D_{train}$, can be transformed into a t_i-dimensional vector

$$\langle \lambda_{i,1}\mathbf{v}_i, \ldots, \lambda_{i,t_i}\mathbf{v}_i \rangle \in \mathbb{F}_i \tag{17.2}$$

where $\lambda_{i,j}\mathbf{v}_i$ represents the j-th component of the value obtained after mapping the instance $\mathbf{x}$ into the feature space. Figure 17.4 gives an illustrative diagram of reducing $\mathbf{v}_1$, $\mathbf{v}_2$, $\mathbf{v}_3$ to features. In particular, each $\mathbf{v}_i$ is reduced to two features $\lambda_{i,1}\mathbf{v}_i$ and $\lambda_{i,2}\mathbf{v}_i$.

Discretisation of Hidden Feature Space

The feature extraction techniques result in mappings $\lambda_{i,j}$ that range over a continuous and potentially infinite domain. Yet, Bayesian network-based abstraction technique relies on the construction of *probability tables*, where each entry associates a set of *distinct* latent feature values with a probability. For this construction to be relevant, we therefore *discretise* each latent feature component into a *finite* set of sub-spaces.

Construction of Bayesian Network Abstraction

The abstraction that we construct primarily represents the *probabilistic distribution* of the set of latent feature values induced by a set $\mathbf{X}$ of test instances. In other words, given an input $\mathbf{x} \in \mathbf{X}$, the abstraction allows us to estimate the probability that $\mathbf{x}$ induces a given combination of values for the latent features that have been learned by the neural network.

Thanks to the layered and acyclic nature of the neural networks that we consider, we can directly characterise the *causal relationship* between the sets of neuron values in various layers *w.r.t.* a series of inputs as well. In other words, given an input $\mathbf{x} \in \mathbf{X}$, one can in principle estimate the conditional probability of each neuron value at layer i *w.r.t.* the probability of every combination of neuron values at layer $i - 1$. By lifting the above relationship from individual neuron values to latent feature intervals, we seek to capture *causal semantic relations* that link the features at each layer: in a layer i, and with an input $\mathbf{x}$, the *probability* that a latent feature valuation belongs to a given interval in the corresponding feature space is *dependent* on probabilities pertained to latent feature intervals at layer $i - 1$.

Preserved Property

First of all, we show that the constructed Bayesian network is an abstraction of the neural network. Given a finite set $\mathbf{X}$ of inputs, an abstraction constructs a set $\mathbf{X}' \supseteq \mathbf{X}$ that generalises $\mathbf{X}$ to more elements [63]. Given an input $\mathbf{x}$ and a Bayesian network $\lambda(\mathbf{X})$, we are able to check the probability of $\mathbf{x}$ on $\lambda(\mathbf{X})$, i.e., $\lambda(\mathbf{X})(\mathbf{x})$. We say that $\mathbf{x}$ is included in $\lambda(\mathbf{X})$ if $\lambda(\mathbf{X})(\mathbf{x}) > 0$. The following lemma suggests that every sample in $\mathbf{X}$ is included in $\lambda(\mathbf{X})$:

Lemma 17.1 *All inputs $\mathbf{x}$ in the dataset $\mathbf{X}$ are included in the $\lambda(\mathbf{X})$ with probability greater than 0.*

Let $\mathbf{X}'$ be the set of inputs that satisfy $\lambda(\mathbf{X})(\mathbf{x}) > 0$. This lemma suggests that $\mathbf{X}' \supset \mathbf{X}$. Therefore, $\lambda(\mathbf{X})$ defines an abstraction of the dataset $\mathbf{X}$. This abstraction also suggests that the abstraction assumption—i.e., inputs that are outliers *w.r.t.* the abstraction are also outliers *w.r.t.* the original neural network—is reasonable because $\mathbf{X}' \supset \mathbf{X}$.

In addition to this simple property, [10, 3] also consider other analysis techniques based on the abstracted Bayesian network.

Exercises

Question 1 Write a PCTL formula to express that the agent will never move into certain states labelled with atomic proposition p. □

Question 2 Write a program to automatically construct a failure process DTMC according to a set of trajectories. □

Question 3 What is the size of model $M^E(\pi, \mathbf{x}_0, C)$, compared with $M^E(\pi, \mathbf{x}_0)$? □

Question 4 Can you identify a test coverage metric that is different from the ones presented in Chap. 14? □

Question 5 What is the relation between reliability and robustness? □

Question 6 Please give a list of assurance techniques with respect to the lifecycle stages of machine learning model. □

Question 7 Can we use the abstracted Bayesian network as described in Chap. 17 for the prediction? If so, what do you think of its accuracy, when comparing with the accuracy of the original network? □

Part V
Appendix: Mathematical Foundations and Competition

The appendix includes contents that serves as the preliminary knowledge of the technical chapters, including probability theory and linear algebra. Only very basic contents are included, and the readers are referred to dedicated textbooks for more in-depth contents. Moreover, we also include some contents for probabilistic graphical model, in addition to the introductory content already provided in Chap. 17. Finally, we also include a competition that is designed for a group of students to learn adversarial attack and training by competing with each other's solutions.

Appendix A
Probability Theory

A.1 Random Variables

In most of our contexts, a random variable is a function that assigns probability values to each of an experiment's (or an event's) outcomes. Intuitively, a random variable has a probability distribution, which represents the likelihood that any of the possible values would occur.

Example A.1 We have a population of students, such that we want to reason about their grades. In this case, we have

- a random variable *Grade*, with the set of possible values $V(Grade) = \{A, B, C\}$, and
- a function $P(Grade)$, which associates a probability with each outcome.

Given $P(X)$ is a probability distribution, we have

$$\sum_{x \in V(X)} P(x) = 1 \tag{A.1}$$

and we may call $P(X)$ the marginal distribution of X, as opposed to the joint and conditional distributions. We will explain these concepts in detail later.

A probability distribution $P(X)$ is a multi-nominal distribution if there are multiple values for the random variable X, i.e., $|V(X)| > 1$. Moreover, if $V(X) = \{false, true\}$ then $P(X)$ is a Bernoulli distribution. Besides, $P(X)$ can be continuous, such that it may take on an uncountable set of values.

Example A.2 The *Grade* random variable in Example A.1 has three possible values and therefore it is associated with a multi-nominal distribution. Also, for a coin tossing example with a single coin that may be either head or tail, the random variable *Coin* is associated with a Bernoulli distribution. Moreover, a real-valued random variable is continuous, even if it only takes values from a real interval.

X. Huang et al., *Machine Learning Safety*, Artificial Intelligence: Foundations, Theory, and Algorithms, https://doi.org/10.1007/978-981-19-6814-3

A random variable can also be multi-dimensional.

Example A.3 For the **iris** example as in Example 1.1, the data instance can be seen as a 4-dimensional continuous variable X, such that it has an underlying data distribution and the dataset is sampled from the data distribution. The label Y can be seen as another random variable.

A.2   Joint and Conditional Distributions

A marginal probability is the probability of an event X occurring, i.e., $P(X)$. It may be thought of as an unconditional probability, as it is not conditioned on any other event.

Example A.4 Assume that we randomly draw a card from a standard deck of playing cards. Let C be the random variable, representing the color of the card drawn. Then, the probability that the card drawn is red is $P(C = red) = 0.5$, or simply $P(red) = 0.5$ if the random variable C is clear from the context.

Example A.5 Given a standard deck of playing cards, the probability that a card drawn is a 4 is $P(four) = 1/13$.

A.2.1   Joint Probability $P(X, Y)$

Joint Probability $P(X, Y)$ is the probability of event X and event Y occurring. It is a statistical measure that calculates the likelihood of two events occurring at the same time.

Example A.6 Given a standard deck of playing cards, the probability that a card is a four and red, i.e., $P(four, red) = 2/52 = 1/26$. Note: there are two red fours in a deck of 52, the 4 of hearts and the 4 of diamonds.

The joint probability can be generalised to work with more than two events.

A.2.2   Conditional Probability $P(X|Y)$

is the probability of event X occurring, given that event Y has already occurred.

Example A.7 Given a standard deck of playing cards, if we know that you have drawn a red card, what is the probability that it is a four? The answer is $P(four|red) = 2/26 = 1/13$, i.e., out of the 26 red cards (given a red card), there are two fours, so $2/26 = 1/13$.

A.3   Independence and Conditional Independence

Without loss of generality, we assume that for the remaining of this section, all random variables are Boolean. Given two Boolean random variables X and Y, we expect that, in general, $P(X|Y)$ is different from $P(X)$, i.e., the fact that Y is true may change our probability over X.

A.3.1   Independence

Sometimes an equality can occur, i.e, $P(X|Y) = P(X)$. i.e., learning that Y occurs does not change our probability of X. In this case, we say event X is independent of event Y, denoted as

$$X \perp Y \tag{A.2}$$

A distribution P satisfies $X \perp Y$ if and only if $P(X, Y) = P(X)P(Y)$.

Example A.8 Consider the joint probability table as shown in Table A.1.
   First of all, we note that

$$\begin{aligned}
P(X = 0) &= 0.32 + 0.8 = 0.4 \\
P(X = 1) &= 0.6 \\
P(Y = 0) &= 0.08 + 0.12 = 0.2 \\
P(Y = 1) &= 0.8
\end{aligned} \tag{A.3}$$

Then, we notice that

$$\begin{aligned}
0.08 &= P(X = 0, Y = 0) = P(X = 0) * P(Y = 0) = 0.4 * 0.2 \\
0.32 &= P(X = 0, Y = 1) = P(X = 0) * P(Y = 1) = 0.4 * 0.8 \\
0.12 &= P(X = 1, Y = 0) = P(X = 1) * P(Y = 0) = 0.6 * 0.2 \\
0.48 &= P(X = 1, Y = 1) = P(X = 1) * P(Y = 1) = 0.6 * 0.8
\end{aligned} \tag{A.4}$$

Table A.1 A simple two variable joint distribution

X	Y	P(X,Y)
0	0	0.08
0	1	0.32
1	0	0.12
1	1	0.48

Table A.2 Another simple
two variable joint distribution

X	Y	P(X,Y)
0	0	0.10
0	1	0.16
1	0	0.64
1	1	0.10

that is,

$$P(X, Y) = P(X) * P(Y) \tag{A.5}$$

which suggests that X and Y are independent.

Example A.9 On the other hand, for the following Table A.2, the two variables X and Y are not independent.

A.3.2 Conditional Independence

While independence is a useful property, we do not often encounter two independent events. A more common situation is when two events X and Y are independent given an additional event Z, denoted as

$$X \perp Y | Z \tag{A.6}$$

A distribution P satisfies $X \perp Y | Z$ if and only if $P(X, Y | Z) = P(X | Z) P(Y | Z)$.

Note that, similar calculation as in Examples A.8 and A.9 can be done to work with conditional independence, with the only changes on replacing the computation of marginal probabilities $P(X)$ and $P(Y)$ with the computation of conditional probabilities $P(X | Z)$ and $P(Y | Z)$.

A.4 Querying Joint Probability Distributions

A joint distribution $P(X_1, \ldots, X_n)$ contains an exponential number 2^n of real probability values and it can be hard to make sense of them. It is desirable that we are able to infer useful information by making queries. In the following, we introduce a few categories of queries.

A.4.1 Probability Queries

are to compute distribution of a subset of random variables, given evidence (i.e., the values) of another subset of random variables. Formally, it is to compute

$$P(X_{i1}, \ldots, X_{ik} | X_{j1} = x_{j1}, \ldots, X_{jl} = x_{jl}) \tag{A.7}$$

where $x_{j1}, \ldots, x_{jl}$ are evidence for the random variables $X_{j1}, \ldots, X_{jl}$, respectively. Moreover, we have $\{X_{i1}, \ldots, X_{ik}\} \cup \{X_{j1}, \ldots, X_{jl}\} \subseteq \{X_1, \ldots, X_n\}$ and $\{X_{i1}, \ldots, X_{ik}\} \cap \{X_{j1}, \ldots, X_{jl}\} = \emptyset$.

A.4.2 Maximum a Posteriori (MAP) Queries

In addition to probability queries which concern the (conditional) probability of the occurrence of events, we may be interested in MAP-style queries, which is to find a joint assignment to some subset of variables that has the highest probability. For simplicity, we let $\{X_1, \ldots, X_k\}$, $\{X_{k+1}, \ldots, X_l\}$, $\{X_{l+1}, \ldots, X_n\}$ be three disjoint subsets of the random variables $\{X_1, \ldots, X_n\}$. Then, the MAP query is of the form

$$
\begin{aligned}
&MAP(X_1, \ldots, X_k | X_{l+1}, \ldots, X_n) \\
&= \arg\max_{x_1, \ldots, x_k} \sum_{x_{k+1}, \ldots, x_l} P(X_1 = x_1, \ldots, X_l = x_l | X_{l+1} = x_{l+1}, \ldots, X_n = x_n)
\end{aligned}
\tag{A.8}
$$

Intuitively, we have the evidence for variables $\{X_{l+1}, \ldots, X_n\}$, and intend to find the joint assignment for $\{X_1, \ldots, X_k\}$. Because we do not have information about $\{X_{k+1}, \ldots, X_l\}$, we marginalise them.

Example A.10 Consider a probability table for variable X_1 as in Table A.3. Then, we have $MAP(X_1) = 1$.

Example A.11 Consider a joint probability table for variables X_1 and X_2 as in Table A.4. Then, we have $MAP(X_2) = 1$.

Table A.3 Probability table for X_1

$X_1 = 0$	$X_1 = 1$
0.4	0.6

Table A.4 Joint probability table for X_1 and X_2

X_1	X_2	$P(X_1, X_2)$
0	0	0.04
0	1	0.36
1	0	0.3
1	1	0.3

Appendix B
Linear Algebra

B.1 Scalars, Vectors, Matrices, Tensors

B.1.1 Scalar

A scalar is a single number. It is represented in lower-case italic, such as x.

Example B.1 We can use $x \in \mathbb{R}$, a real-valued scalar, to define the slope of the line. Moreover, we can use $x \in \mathbb{N}$, a natural number scalar, to define the number of units.

B.1.2 Vector

A vector is an array of numbers arranged in order. Each number in a vector is identified with an index. Vectors are represented as lower-case bold, such as $\mathbf{x}$. If x is n-dimensional and each element of $\mathbf{x}$ is a real number, we write $\mathbf{x} \in \mathbb{R}^n$.

Geometrically, we think of vectors as points in space such that each element of a vector gives coordinate along an axis. This is the same as we consider a data instance—usually represented as a vector—as a point in high-dimensional space.

B.1.3 Matrix

A matrix is a 2-D array of numbers, such that each element is identified by two indices. Matrices are represented as bold typeface $\mathbf{A}$. We usually identify each element of $\mathbf{A}$ with the subscripts, i.e., use $\mathbf{A}_{i,j}$ to denote the element on the i-th row and j-th column. We may also write $\mathbf{A}[i :]$ for the i-th row of $\mathbf{A}$ and $\mathbf{A}[: j]$ the

j-th column of $\mathbf{A}$. Similar as the vectors, we may write $\mathbf{A} \in \mathbb{R}^{m \times n}$ if $\mathbf{A}$ has m rows and n columns and each element is a real number.

B.1.4 Tensor

It is very likely in a machine learning context that, we may need an array with more than two axes. We call a multi-dimensional array arranged on a regular grid with variable number of axes as a tensor. Tensors are also represented as bold typeface $\mathbf{A}$, and we may use the subscripts to identify its element, for example $\mathbf{A}_{i,j,k}$ to denote the element in a three-dimensional tensor. We remark that, a tensor may include more involved properties that a multi-dimensional array does not have, but we omit the details for the simplicity of the explanations.

B.2 Matrix Operations

We briefly review a few frequently used matrix operations.

B.2.1 Transpose of a Matrix

The transpose of a matrix is an operator which flips a matrix over its diagonal. Given a matrix $\mathbf{A}$, we define its transpose $\mathbf{A}$ as that

$$\mathbf{A}_{ij}^{T} = \mathbf{A}_{ji} \text{ for all } i, j \tag{B.1}$$

B.2.2 Linear Transformation

A linear transformation is a function from one vector space to another that respects the underlying (linear) structure of each vector space. Linear transformations can be represented by matrices. If T is a linear transformation mapping from $\mathbb{R}^{n}$ to $\mathbb{R}^{m}$, and $\mathbf{x}$ is a column vector with n entries, then

$$T(\mathbf{x}) = \mathbf{A}\mathbf{x} \tag{B.2}$$

for some $m \times n$ matrix $\mathbf{A}$. Also, $\mathbf{A}$ is called transformation matrix of T.

B.2.3   Identity Matrix

The identity matrix of size n is the $n \times n$ square matrix with ones on the diagonal and zeros elsewhere. We use $\mathbf{I}$ to denote an identity matrix.

B.2.4   Matrix Inverse

An $n \times n$ matrix $\mathbf{A}$ is invertible if there exists another $n \times n$ matrix $\mathbf{A}$ such that

$$\mathbf{AB} = \mathbf{BA} = \mathbf{I} \tag{B.3}$$

We write $\mathbf{B}$ as the inverse of $\mathbf{A}$, denoted as $\mathbf{A}^{-1} = \mathbf{B}$,

B.3   Norms

Usually, a distance function is employed to compare data instances. Ideally, such a distance should reflect perceptual similarity between data instances, comparable to e.g., human perception for image classification networks. A distance metric should satisfy a few axioms which are usually needed for defining a metric space:

- $||\mathbf{x}|| \geq 0$ (non-negativity),
- $||\mathbf{x} - \mathbf{y}|| = 0$ implies that $\mathbf{x} = \mathbf{y}$ (identity of indiscernibles),
- $||\mathbf{x} - \mathbf{y}|| = ||\mathbf{y} - \mathbf{x}||$ (symmetry),
- $||\mathbf{x} - \mathbf{y}|| + ||\mathbf{y} - \mathbf{z}|| \geq ||\mathbf{x} - \mathbf{z}||$ (triangle inequality).

In practise, L_p-norm distances are used, including

- L_1 (Manhattan distance):

$$||\mathbf{x}||_1 = \sum_{i=1}^{n} |x_i| \tag{B.4}$$

- L_2 (Euclidean distance):

$$||\mathbf{x}||_2 = \sqrt{\sum_{i=1}^{n} x_i^2} \tag{B.5}$$

- L_∞ (Chebyshev distance):

$$||\mathbf{x}||_\infty = \max_i |x_i| \tag{B.6}$$

Moreover, we also consider L_0-norm as $||\mathbf{x}||_0 = |\{x_i \mid x_i \neq 0, i = 1..n\}|$, i.e., the number of non-zero elements. Note that, L_0-norm does not satisfy the triangle inequality. In addition to these, there exist other distance metrics such as Fréchet Inception Distance [64]. In addition, Frobenius norm of a matrix is represented by $|| \cdot ||_{Fr}$.

Given a data instance x and a distance metric L_p, the *neighbourhood* of $\mathbf{x}$ is defined as follows.

Definition B.1 (d-Neighbourhood) Given a data instance $\mathbf{x}$, a distance function L_p, and a distance d, we define the *d-neighbourhood* $\eta(\mathbf{x}, L_p, d)$ of $\mathbf{x}$ w.r.t. L_p as

$$\eta(\mathbf{x}, L_p, d) = \{\hat{\mathbf{x}} \mid ||\hat{\mathbf{x}} - \mathbf{x}||_p \leq d\}, \tag{B.7}$$

the set of data instances whose distance to $\mathbf{x}$ is no greater than d with respect to L_p.

B.4   Variance, Covariance, and Covariance Matrix

B.4.1   Variance

Variance measures the variation of a single random variable. Formally,

$$Var(X) = \mathbb{E}[(X - \overline{X})^2] \tag{B.8}$$

where $\overline{X} = \mathbb{E}[X]$ is the mean. If we are working with a finite set D of data samples, we have

$$\overline{X} = \frac{1}{|D|} \sum_{\mathbf{x} \in D} \mathbf{x} \qquad\qquad Var(X) = \frac{1}{|D|} \sum_{\mathbf{x} \in D} (\mathbf{x} - \bar{X})^2 \tag{B.9}$$

where $|D|$ denotes the number of samples in D.

B.4.2   Covariance

Covariance based on variance, measures how much two random variables vary together. Formally,

$$Cov(X, Y) = \mathbb{E}[(X - \mathbb{E}[X])(Y - \mathbb{E}[Y])] \tag{B.10}$$

If working with a dataset D of $(\mathbf{x}, \mathbf{y})$ pairs, it is

$$Cov(X, Y) = \frac{1}{|D|} \sum_{(\mathbf{x},\mathbf{y}) \in D} (\mathbf{x} - \bar{X})(\mathbf{y} - \bar{Y}) \tag{B.11}$$

B.4.3 Covariance Matrix

Generalising the above, consider a column vector $\mathbf{X} = (X_1, \ldots, X_n)^T$ of random variables, we can have the covariance matrix as follows:

$$\mathbf{K} = \begin{bmatrix} Cov(X_1, X_1) & Cov(X_1, X_2) & Cov(X_1, X_3) & \ldots & Cov(X_1, X_n) \\ Cov(X_2, X_1) & Cov(X_2, X_2) & Cov(X_2, X_3) & \ldots & Cov(X_2, X_n) \\ \ldots\ldots\ldots\ldots\ldots\ldots\ldots\ldots\ldots\ldots \\ Cov(X_n, X_1) & Cov(X_n, X_2) & Cov(X_n, X_3) & \ldots & Cov(X_n, X_n) \end{bmatrix} \tag{B.12}$$

Appendix C
Probabilistic Graph Models

C.1 I-Maps

This section studies the conditional independences in G.

C.1.1 Naive Bayes and Joint Probability

First of all, we explain the difference between usual classification and Naive Bayes classifier. As shown in Fig. C.1, usual classification can be represented as a graphical model G where the class label Y receives incoming connections from the feature variables $X_1, \ldots, X_n$. Therefore, Y has a conditional probabilistic table $P(Y|X_1, \ldots, X_n)$ and the feature variables X_i has a marginal probability table $P(X_i)$. However, for Naive Bayes classifier, it can be represented as another graphical model G' where the label Y has outgoing arrows to the feature variables $X_1, \ldots, X_n$. Therefore, each feature variable X_i has a conditional probability table $P(X_i|Y)$ and the class label has a marginal probability table $P(Y)$. As we explained earlier, for naive Bayes, we have

$$P(X_1, \ldots, X_n, Y) = P(Y) \prod_{i \in \{1..n\}} P(X_i|Y) \tag{C.1}$$

That is, the joint probability is the product of probability tables for the nodes on the graph. Generalising the case for Naive Bayes classifier, we conjecture that for each probabilistic graphical model $G = (\mathcal{V}, \mathcal{E}, P)$ represent the joint probability distribution between random variables, we may have

$$P(\mathcal{V}) = \prod_{V \in \mathcal{V}} P(V|Pa(V)) \tag{C.2}$$

X. Huang et al., *Machine Learning Safety*, Artificial Intelligence: Foundations, Theory, and Algorithms, https://doi.org/10.1007/978-981-19-6814-3

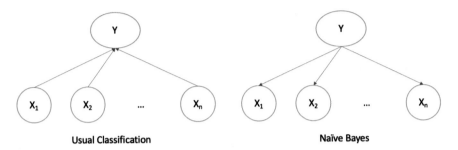

Fig. C.1 Naive Bayes as a probabilistic graphical model

where $Pa(V)$ is the set of parent nodes of node V. Note that, we let $P(V|\emptyset) = P(V)$ when the node V does not have incoming edges.

C.1.2   Independencies in a Distribution

Before proceeding to the conditional probabilities in G, we may need to know how to find conditional independencies from a joint probability. For this, Sect. A.3 has presented a few examples on how to do the calculation.

C.1.3   Markov Assumption

In addition to the notation $Pa(X)$ for the set of parents of a node X, we need $NonDesc(X)$ for the set of non-descendents of X. We have that

> each random variable X is independent of its non-descendents, given its parents, i.e.,
>
> $$X \perp NonDesc(X)|Pa(X) \tag{C.3}$$

Intuitively, parents of a variable shield it from probabilistic influence. Once values of parents are known, no influence of ancestors can be made. On the other hand, information about descendants can change beliefs about a node .

For the running example in Fig. 17.2, we have the following local conditional independences that can be read directly from the graph:

$$I(G) = \{ \ (Fog \perp Detected, Radar, Away, Stopped | Camera),$$
$$(Camera \perp Radar, Away),$$
$$(Radar \perp Camera, Fog),$$
$$(Detected \perp Away, Fog | Camera, Radar), \tag{C.4}$$
$$(Away \perp Camera, Fog, Detected | Radar),$$
$$(Stopped \perp Fog, Camera, Radar | Detected, Away)\}$$

Intuitively, to understand $(Fog \perp Detected, Radar, Away, Stopped | Camera)$, we note that, once we are able to observe the camera's result, the weather conditions such as Fog can be directly obtained and are independent from other random variables. Moreover, to understand $(Camera \perp Radar, Away)$, we note that camera's precision—as a quality of hardware—is absolutely independent of both radar's result and whether or not the pedestrian is away. Other (conditional) independencies can be explained in a similar way.

We remark that, the conditional independencies obtained through this way are local ones, and do not necessarily include all conditional independencies that can be inferred from the graph. In Sect. C.3, we will introduce a comprehensive method—d-separation—that is able to infer all possible conditional independencies from a graph.

C.1.4 I-Map of Graph and Factorisation of Joint Distribution

Let G be a graph associated with a set of independencies $I(G)$, and P a probability distribution with a set of independencies $I(P)$. Note that, $I(P)$ can be obtained by the way shown in Sect. C.1.2. Then, we define

Definition C.1 G is an I-Map of P if $I(G) \subseteq I(P)$.

By this definition, I-Map requires that a joint distribution P can have more independencies than the graph G, but graph G cannot mislead by containing independencies that do not exist in P.

Example C.1 Consider the joint probability table P as in Table A.1, and the three graphs in Fig. C.2. we note that $I(P) = \{X \perp Y\}$, $I(G_0) = \{X \perp Y\}$, and $I(G_1) = I(G_2) = \emptyset$. Therefore, all three graphs G_0, G_1, G_2 are I-maps of P. On the other hand, if consider the joint probability table P as in Table A.2, we have that $I(P) = \emptyset$. Then, while G_1 and G_2 are still I-maps of P, G_0 is not any more.

In the following, we introduce the relationship between I-map and factorisation. Factorization is to write a mathematical object as a product of several, usually smaller or simpler, objects of the same kind. For example, in the Naive Bayes classifier, the joint distribution $P(X_1, \ldots, X_n, Y)$ is rewritten as the production of $P(Y)$ and conditional probabilities $P(X_i | Y)$, where $P(Y)$ and $P(X_i | Y)$ are much smaller than $P(X_1, \ldots, X_n, Y)$. For the graphical model in general, we have

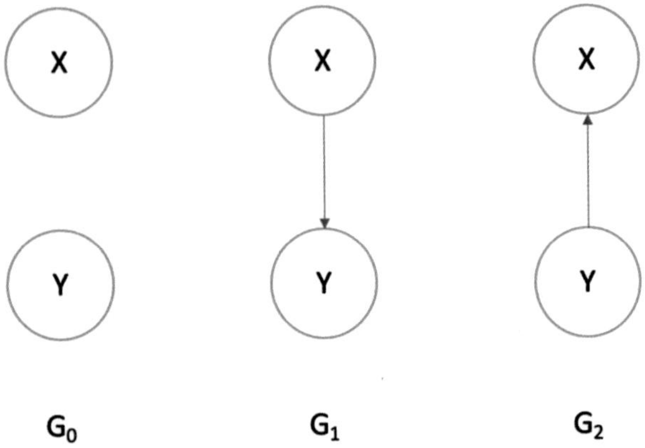

Fig. C.2 Three simple, two-node graphs

G_0 G_1 G_2

Theorem C.1 *If G is an I-map of P, then*

$$P(X_1, \ldots, X_n) = \prod_{i=1}^{n} P(X_i | Pa(X_i)) \qquad (C.5)$$

This justifies our conjecture at Eq. (C.2).

C.1.5   Perfect Map

Similar as I-map, we may define D-map, which requires that $I(P) \subseteq I(G)$. The intersection of I-map and D-map leads to perfect map, which requires that the conditional independencies in G and P. Interestingly, but not surprisingly, not all distributions P over a given set of variables can be represented as a perfect map.

C.2   Reasoning Patterns

As explained earlier, the construction of graphical models will enable our reasoning about a more complex relation between a set of random variables. In this section, we introduce a few typical reasoning patterns.

C.2.1   Causal Reasoning

Causal reasoning considers how the changes of up-stream variables may affect the values of the down-stream variables. For our running example as in Fig. 17.3, it is for an engineer to concern about how likely the car will stop given e.g., the quality of sensors (i.e., camera and/or radar).

A typical process is to first compute a marginal probability such as

$$P(s_0) \tag{C.6}$$

which is the probability of the car does not stop. This can be computed by having

$$
\begin{aligned}
P(s_0) &= \sum_{C,R,F,D,A} P(C, R, F, D, A, S = s_0) \\
&= \sum_{i=0}^{1} \sum_{j=0}^{1} \sum_{k=0}^{1} \sum_{l=0}^{1} \sum_{m=0}^{1} P(C = c_i, R = r_j, F = f_k, D = d_l, A = a_m, S = s_0) \\
&\approx 0.62
\end{aligned}
\tag{C.7}
$$

Then, we may consider a conditional probability such as

$$P(s_0|c_1) = \frac{P(s_0, c_1)}{P(c_1)} \approx 0.51 \tag{C.8}$$

which says that if we know that the camera captured the pedestrian, the probability of not stopped decreased to 0.51. Similarly, if consider radar, we have

$$P(s_0|r_1) = \frac{P(s_0, c_1)}{P(r_1)} \approx 0.44 \tag{C.9}$$

If both the camera and the radar are considered, we have

$$P(s_0|c_1, r_1) = \frac{P(s_0, c_1, r_1)}{P(c_1, r_1)} \approx 0.31 \tag{C.10}$$

C.2.2   Evidential Reasoning

A driver may want to know, subject to her own experience on e.g., car stopping and foggy weather, about the quality of the sensors. For example, first of all, she may have the following statistics from the vendor of the camera:

$$P(c_1) = 0.4 \tag{C.11}$$

After observing that the car stopped, she might infer as follows, which shows that the probability increased to 0.51.

$$P(c_1|s_1) = \frac{P(c_1, s_1)}{P(s_1)} \approx 0.51 \qquad (C.12)$$

Intuitively, this may suggest that the specific camera installed on this car may perform above average.

C.2.3   Inter-Causal Reasoning

In addition to the above, it might be interested to understand how the causes of an event may affect each other. For example, consider the following

$$P(r_1|d_1) \approx 0.83 \qquad (C.13)$$

which suggests that once we know that the pedestrian is detected, the chance of radar captured the pedestrian is high, i.e., the quality of the radar is good. However, by having the following

$$P(r_1|d_1, c_1) \approx 0.75 \qquad (C.14)$$

which is lower than $P(r_1|d_1)$, it suggests that the quality of the radar may not be as optimistic as it seems when only observing the detection result. The quality of the camera may also contribute well to the excellent detection result.

C.2.4   Practice

First of all, we install a software package that can support the inference of probabilistic graphical models:

```
$ conda install pomegranate
```

To work with the package, we create a script **pedestrian_detection.txt**. In the script, first of all, we import the package:

```
from pomegranate import *
```

Then, we encode a graphical model

```
camera = DiscreteDistribution({'0': 0.6, '1': 0.4})
radar = DiscreteDistribution({'0': 0.5, '1': 0.5})
fog = ConditionalProbabilityTable(
        [['0', '0', 0.5],
         ['0', '1', 0.5],
         ['1', '0', 0.3],
         ['1', '1', 0.7]], [camera])
```

```
 8 away = ConditionalProbabilityTable(
 9         [['0', '0', 0.2],
10          ['0', '1', 0.8],
11          ['1', '0', 0.6],
12          ['1', '1', 0.4]], [radar])
13 detected = ConditionalProbabilityTable(
14         [['0', '0', '0', 1.0],
15          ['0', '0', '1', 0.0],
16          ['0', '1', '0', 0.6],
17          ['0', '1', '1', 0.4],
18          ['1', '0', '0', 0.7],
19          ['1', '0', '1', 0.3],
20          ['1', '1', '0', 0.1],
21          ['1', '1', '1', 0.9]], [camera, radar])
22 stopped = ConditionalProbabilityTable(
23         [['0', '0', '0', 0.6],
24          ['0', '0', '1', 0.4],
25          ['0', '1', '0', 0.9],
26          ['0', '1', '1', 0.1],
27          ['1', '0', '0', 0.1],
28          ['1', '0', '1', 0.9],
29          ['1', '1', '0', 0.5],
30          ['1', '1', '1', 0.5]], [detected, away])
31
32 s1 = Node(camera, name="camera")
33 s2 = Node(radar, name="radar")
34 s3 = Node(fog, name="fog")
35 s4 = Node(away, name="away")
36 s5 = Node(detected, name="detected")
37 s6 = Node(stopped, name="stopped")
38
39 model = BayesianNetwork("Pedestrian Detection Problem")
40 model.add_states(s1, s2, s3, s4, s5, s6)
41 model.add_edge(s1, s3)
42 model.add_edge(s1, s5)
43 model.add_edge(s2, s4)
44 model.add_edge(s2, s5)
45 model.add_edge(s5, s6)
46 model.add_edge(s4, s6)
47 model.bake()
```

After the above, we can start computing the probability values:

```
 1 #### P(s0,c1)
 2 query = ['1', None, None, None, None, '0']
 3 ps0c1 = 0
 4 for j1 in range(2):
 5     for j2 in range(2):
 6         for j3 in range(2):
 7             for j4 in range(2):
 8                 ps0c1 += model.probability([['1', str(j1), str(j2
    ), str(j3), str(j4), '0']])
 9 print("the probability of the car does not stop but the camera
    captured the pedestrian P(s0,c1): %s\n"%(ps0c1))
10
```

```
11  ### P(c1)
12  query = ['1', None, None, None, None, None]
13  pc1 = 0
14  for j1 in range(2):
15      for j2 in range(2):
16          for j3 in range(2):
17              for j4 in range(2):
18                  for j5 in range(2):
19                      pc1 += model.probability([['1', str(j1), str(
      j2), str(j3), str(j4), str(j5)]])
20  print("the probability of camera captured the pedestrian P(c1): %
      s\n"%pc1)
21
22  #### P(s0|c1)
23  print("the conditional probability of the car does not stop when
      the camera captured the pedestrian P(s0|c1)): %s\n"%(ps0c1/
      pc1))
```

C.3   D-Separation

From Sect. C.1, we know that a graph structure G encodes a set of conditional independence assumptions $I(G)$ and we can read a set of independencies directly according to the Markov assumption. However, we may be interested to infer all possible conditional independence from a graph G.

Definition C.2 D-separation is a procedure d-sep$_G(X \perp Y|Z)$ that, given a graph G and three sets X, Y, and Z of nodes in G, returns Yes or No, such that d-sep$_G(X \perp Y|Z) = Yes$ iff $(X \perp Y|Z)$ follows from $I(G)$.

First of all, we note that if X and Y are connected directly, they are co-related regardless of any evidence about any other variables. So, in the following, we consider X and Y that are not directly connected.

C.3.1   Four Local Triplets

Before considering more complex cases, we consider four local cases where X and Y are indirectly connected with another variable Z in the middle. The four possible cases are shown in Fig. C.3.

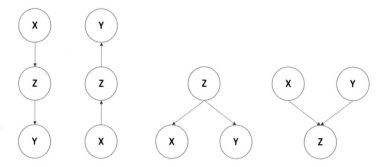

Fig. C.3 Four patterns

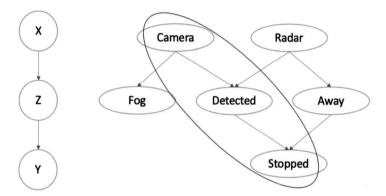

Fig. C.4 Indirect causal effect

C.3.2   Indirect Causal Effect $X \rightarrow Z \rightarrow Y$

Cause X cannot influence effect Y if Z is observed, i.e., observed Z blocks the influence of X over Y. For the running example, as shown in Fig. C.4,

C.3.3   Indirect Evidential Effect $Y \rightarrow Z \rightarrow X$

Similarly, evidence X cannot influence the cause Y if Z is observed.

C.3.4   Common Cause $X \leftarrow Z \rightarrow Y$

Once Z is observed, one of the effects cannot influence the other. Figure C.5 presents a case of common cause in our running example.

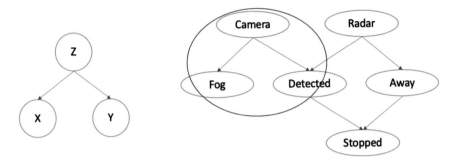

Fig. C.5 Common Cause

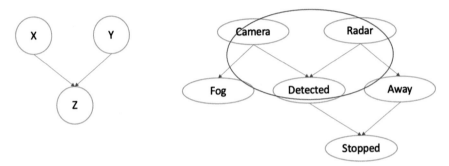

Fig. C.6 Common effect

C.3.5 *Common Effect X → Z ← Y*

Unlike the above three cases where observing the middle variable Z blocks the influence, the case of common effect is on the opposite, i.e., the influence is blocked when the common effect Z and its descendants are not observed. Figure C.6 presents a case of common effect in our running example.

C.3.6 *General Case: Active Trail and D-Separation*

Summarising the above discussion on local canonical cases (or v-structures), we have that

- Causal trail, $X → Z → Y$, is active if and only if Z is not observed.
- Evidential trail, $X ← Z ← Y$, is active if and only if Z is not observed.
- Common cause, $X ← Z → Y$, is active if and only if Z is not observed.
- Common effect, $X → Z ← Y$, is active if and only if either Z or one of its descendants is observed.

Now we can consider the general case of d-sep$_G(X \perp Y | Z)$, where X, Y and Z may not be connected to each other. Actually, the graph can be large and the variables may be far away from each other. Fortunately, as we will introduce below, any complex case can be broken into repetitions of the local canonical cases.

The D-separation algorithm is given in Algorithm 12. It collects all paths between X and Y, without considering the directions of the edges. Then, as long as no paths are active given the observations $\{Z_1, \ldots, Z_k\}$, the two variables X and Y are conditionally independent.

Algorithm 12: *d-sep*$_G(X, Y, \{Z_1, \ldots, Z_k\})$, where X, Y, Z_i are nodes on a graph G

---

1 $Path$ = all (undirected) paths from X to Y
2 **for** $path$ in $Path$ **do**
3   $\quad$ $isActive(path, \{Z_1, \ldots, Z_k\})$
4 **end**
5 **return** $X \not\perp Y | \{Z_1, \ldots, Z_k\}$ **return** $X \perp Y | \{Z_1, \ldots, Z_k\}$

Now, we need to determine whether a path is active or not, i.e., *isActive*(path, $\{Z_1, \ldots, Z_k\}$), which can be done with Algorithm 13. Simply speaking, it requires all local triplets on the path to be inactive to make the path inactive.

Algorithm 13: *isActive*$(path, \{Z_1, \ldots, Z_k\})$, where $path$ is a path on the graph and Z_i are nodes on G

---

1 Let $path = X_1, \ldots, X_k$
2 **for** all triplets X_{i-1}, X_i, X_{i+1} on path **do**
3   $\quad$ (X_{i-1}, X_i, X_{i+1}) is inactive
4 **end**
5 **return** False **return** True

C.3.7 I-Equivalence

Conditional independence assertions can be the same with different graphical structures, and I-equivalence is to capture such equivalence relation.

Definition C.3 Two graphs G_1 and G_2 are I-equivalent if $I(G_1) = I(G_2)$.

Let skeleton of a graph G be an undirected graph with an edge for every edge in G. We have that, if two graphs have the same set of skeletons and v-structures then they are I-equivalent.

C.4   Structure Learning

There are mainly two approaches to acquire a graphical model. The first is through knowledge engineering, where the graphical model is constructed by hand with expert's help. The second is through machine learning, by which the graphical model is learned from a set of instances.

First of all, the following is the problem statement for structure learning:

> Assume dataset D is generated i.i.d. from distribution $P^*(X)$, and $P^*(X)$ is induced by an underlying graph G^*. The goal is to construct a graphical model G such that it is as close as possible to G^*.

Considering that there may be many I-maps for P^* and we cannot distinguish them from D, we know that G^* is not identifiable. Nevertheless, in most cases, it is sufficient to recover G^*'s equivalence class.

C.4.1   Criteria of Structure Learning

While our goal is to recover G^*'s equivalence class, this goal is nontrivial. One of the key issues is that the data instances sampled from the distribution $P^*(X)$ may be noisy. Therefore, we need to make decisions about including edges we are less sure about. Actually, too few edges means the possibility of missing out on dependencies, and too many edges means the possibility of spurious dependencies.

Example C.2   In a coin tossing example, we have two coins X and Y, who are tossed independently. Assume that they are tossed 100 times, with the statistics shown in Table C.1. Are X and Y independent? Actually, if we follow the exact computation, they are dependent. But we suspect that they should be independent, because the probability of getting exactly 25 in each category is small (approx. 1 in 1,000).

Actually, in a conditional probability table, the number of entries grows exponentially with the of parent nodes. Therefore, the cost of adding a parent node can be very large. According to this, it is better to obtain a sparser, simpler structure. Actually, we can sometimes learn a better model by learning a model with fewer edges even if it does not represent the true distribution.

Table C.1   A simple coin tossing example

X	Y	Times
Head	Head	27
Head	Tail	22
Tail	Head	25
Tail	Head	26

C.4.2 Overview of Structure Learning Algorithms

Basically, there are two classes of structural learning algorithms:

- Constraint-based algorithms
- Score-based algorithms

Constraint-based algorithms are to find a graphical model whose implied independence constraints match those found in the data. On the other hand, score-based algorithms are to find a graphical model that can represent distributions that match the data (i.e., could have generated the data).

In the following, we introduce two different directions of structural learning, by the consideration of learning local relations and global structure, respectively.

C.4.3 Local: Independence Tests

For testing the independence between variables, we need measures of "deviance-from-independence" and rules for accepting/rejecting hypothesis of independence. For the measures, we may consider e.g., mutual Information (K-L divergence) between joint and product of marginals, by having

$$d_I(D) = \frac{1}{|D|} \sum_{x_i, x_j} P(x_i, x_j) \log \frac{P(x_i, x_j)}{P(x_i)P(x_j)} \qquad (C.15)$$

for any two variables X_i and X_j. Theoretically, $d_I(D) = 0$ if X_i and X_j are independent, and $d_I(D) > 0$ otherwise. In addition to $d_I(D)$, there may be other means to define the measures, such as Pearson's Chi-squared test.

Based on the measure, we can define the acceptance rule as

$$R_{d,t}(D) = \begin{cases} \text{Accept} & d(D) \leq t \\ \text{Reject} & d(D) > t \end{cases} \qquad (C.16)$$

where t is a pre-specified threshold. We remark that false rejection probability due to choice of t is its p-value (refer to hypothesis testing). Alternatively, we may take Chow-Liu algorithm to construct a tree-like graphical model as follows:

1. find maximum weight spanning tree based on the values we compute in Eq. (C.15); there are existing algorithms for this purpose, such as the Kruskal's algorithm and the Prim's algorithm.
2. pick a root node and assign edge directions.

C.4.4   Global: Structure Scoring

Similarly as the local case, we need measures to evaluate the goodness of a graphical model and rules for accepting/rejecting hypothesis of goodness. There are different ways to define measures, including

- Log-likelihood Score for G with n variables

$$Score_L(G, D) = \sum_D \sum_{i=1}^n \log P(x_i | pa(x_i)) \qquad (C.17)$$

which can be seen as the loss of using graph G to predict D.
- Bayesian Score

$$Score_B(G, D) = \log P(D|G) + \log P(G) \qquad (C.18)$$

which, in addition to the log-likelihood score $\log P(D|G)$, it also consider the prior $P(G)$.
- Bayes score with penalty term

$$Score_{BIC}(G, D) = L(G, D) - \frac{\log |D|}{2} ||G|| \qquad (C.19)$$

where $L(G, D)$ is the loss of using G to predict D, and $||G||$ is the complexity of the graph G.

Once defined the scores, the structure learning is to search for a graphical structure with the highest score. It is known that finding the optimal one among those structures with at most k parents is NP-hard for $k > 1$. To deal with the high complexity, there are multiple methods including e.g.,

- Greedy search
- Greedy search with restarts
- MCMC methods

For the greedy search, it repeatedly does the following:

1. score all possible single changes, and
2. select the best change to apply if there are any changes that lead to better performance than the existing structure.

C.4.5   Practice

The following code uses the **pomegranate** package to automatically learn structures from a simple dataset.

```python
from pomegranate import BayesianNetwork
import seaborn, time, numpy, matplotlib
seaborn.set_style('whitegrid')

import pandas as pd
X = pd.DataFrame({'1':[0,0,0,1,0], '2':[0,0,1,0,0], '3'
    :[1,1,0,0,1], '4':[0,1,0,1,1]})
X = X.to_numpy()

tic = time.time()
model = BayesianNetwork.from_samples(X)
t = time.time() - tic
p = model.log_probability(X).sum()
print("Greedy")
print("Time (s): ", t)
print("P(D|M): ", p)
model.plot()

tic = time.time()
model = BayesianNetwork.from_samples(X, algorithm='exact-dp')
t = time.time() - tic
p = model.log_probability(X).sum()
print("exact-dp")
print("Time (s): ", t)
print("P(D|M): ", p)
model.plot()

tic = time.time()
model = BayesianNetwork.from_samples(X, algorithm='exact')
t = time.time() - tic
p = model.log_probability(X).sum()
print("exact")
print("Time (s): ", t)
print("P(D|M): ", p)
model.plot()
```

Appendix D
Competition: Resilience to Adversarial Attack

This is a student competition to address two key issues in modern deep learning, i.e.,

O1   how to find better adversarial attacks, and
O2   how to train a deep learning model with better robustness to the adversarial attacks.

We provide a template code (**Competition/Competition.py**), where there are two code blocks corresponding to the training and the attack, respectively. The two code blocks are filled with the simplest implementations representing the baseline methods, and the participators are expected to replace the baseline methods with their own implementations, in order to achieve better performance regarding the above O1 and O2.

D.1   Submissions

In the end, we will collect submissions from the students and rank them according to a pre-specified metric taking into consideration both O1 and O2. Assume that we have n students participating in this competition, and we have a set S of submissions.

> Every student with student number i will submit a package i**.zip**, which includes two files:
>
> 1. i**.pt**, which is the file to save the trained model, and
> 2. competition_i**.py**, which is your script after updating the two code blocks in **Competition.py** with your implementations.
>
> NB: Please carefully follow the naming convention as indicated above, and we will not accept submissions which do not follow the naming convention.

© The Author(s), under exclusive license to Springer Nature Singapore Pte Ltd. 2023   295
X. Huang et al., *Machine Learning Safety*, Artificial Intelligence: Foundations,
Theory, and Algorithms, https://doi.org/10.1007/978-981-19-6814-3

D.2   Source Code

The template source code of the competition is available at

> https://github.com/xiaoweih/
> AISafetyLectureNotes/tree/main/Competition

In the following, we will explain each part of the code.

D.2.1   Load Packages

First of all, the following code piece imports a few packages that are needed.

```
import numpy as np
import pandas as pd
import torch
import torch.nn as nn
import torch.nn.functional as F
from torch.utils.data import Dataset, DataLoader
import torch.optim as optim
import torchvision
from torchvision import transforms
from torch.autograd import Variable
import argparse
import time
import copy
```

Note: You can add necessary packages for your implementation.

D.2.2   Define Competition ID

The below line of code defines the student number. By replacing it with your own student number, it will automatically output the file i.**pt** once you trained a model.

```
# input id
id_ = 1000
```

D.2.3   Set Training Parameters

The following is to set the hyper-parameters for training. It considers e.g., batch size, number of epochs, whether to use CUDA, learning rate, and random seed. You may change them if needed.

```
# setup training parameters
parser = argparse.ArgumentParser(description='PyTorch MNIST
    Training')
parser.add_argument('--batch-size', type=int, default=128,
    metavar='N',
                    help='input batch size for training (default:
    128)')
parser.add_argument('--test-batch-size', type=int, default=128,
    metavar='N',
                    help='input batch size for testing (default:
    128)')
parser.add_argument('--epochs', type=int, default=10, metavar='N'
    ,
                    help='number of epochs to train')
parser.add_argument('--lr', type=float, default=0.01, metavar='LR
    ',
                    help='learning rate')
parser.add_argument('--no-cuda', action='store_true', default=
    False,
                    help='disables CUDA training')
parser.add_argument('--seed', type=int, default=1, metavar='S',
                    help='random seed (default: 1)')
args = parser.parse_args(args=[])
```

D.2.4   Toggle GPU/CPU

Depending on whether you have GPU in your computer, you may toggle between devices with the below code. Just to remark that, for this competition, the usual CPU is sufficient and a GPU is not needed.

```
# judge cuda is available or not
use_cuda = not args.no_cuda and torch.cuda.is_available()
#device = torch.device("cuda" if use_cuda else "cpu")
device = torch.device("cpu")

torch.manual_seed(args.seed)
kwargs = {'num_workers': 1, 'pin_memory': True} if use_cuda else
    {}
```

D.2.5   Loading Dataset and Define Network Structure

In this competition, we use the same dataset (FashionMNIST) and the same network architecture. The following code specify how to load dataset and how to construct a 3-layer neural network. Please do not change this part of code.

```
1  ###################################################don't change
       the below code
       #################################################
2
3  train_set = torchvision.datasets.FashionMNIST(root='data', train=
       True, download=True, transform=transforms.Compose([transforms
       .ToTensor()]))
4  train_loader = DataLoader(train_set, batch_size=args.batch_size,
       shuffle=True)
5
6  test_set = torchvision.datasets.FashionMNIST(root='data', train=
       False, download=True, transform=transforms.Compose([
       transforms.ToTensor()]))
7  test_loader = DataLoader(test_set, batch_size=args.batch_size,
       shuffle=True)
8
9  # define fully connected network
10 class Net(nn.Module):
11     def __init__(self):
12         super(Net, self).__init__()
13         self.fc1 = nn.Linear(28*28, 128)
14         self.fc2 = nn.Linear(128, 64)
15         self.fc3 = nn.Linear(64, 32)
16         self.fc4 = nn.Linear(32, 10)
17
18     def forward(self, x):
19         x = self.fc1(x)
20         x = F.relu(x)
21         x = self.fc2(x)
22         x = F.relu(x)
23         x = self.fc3(x)
24         x = F.relu(x)
25         x = self.fc4(x)
26         output = F.log_softmax(x, dim=1)
27         return output
28
29 ###################################################end of "don't
       change the below code"
       #################################################
```

D.2.6   Adversarial Attack

The part is the place needing your implementation, for O1. In the template code, it includes a baseline method which uses random sampling to find adversarial attacks. You can only replace the middle part of the function with your own implementation (as indicated in the code), and are not allowed to change others.

```
1  'generate adversarial data, you can define your adversarial
       method'
2  def adv_attack(model, X, y, device):
```

```
 3    X_adv = Variable(X.data)
 4
 5    ################################################Note: below is
      the place you need to edit to implement your own attack
      algorithm
 6    ############################################
 7
 8    random_noise = torch.FloatTensor(*X_adv.shape).uniform_(-0.1,
      0.1).to(device)
 9    X_adv = Variable(X_adv.data + random_noise)
10
11    ############################################ end of attack
      method
12    ############################################
13
14    return X_adv
```

D.2.7 Evaluation Functions

Below are two supplementary functions that return loss and accuracy over test
dataset and adversarially attacked test dataset, respectively. We note that the function
adv_attack is used in the second function. You are not allowed to change these two
functions.

```
 1  'predict function'
 2  def eval_test(model, device, test_loader):
 3      model.eval()
 4      test_loss = 0
 5      correct = 0
 6      with torch.no_grad():
 7          for data, target in test_loader:
 8              data, target = data.to(device), target.to(device)
 9              data = data.view(data.size(0),28*28)
10              output = model(data)
11              test_loss += F.nll_loss(output, target, size_average=
      False).item()
12              pred = output.max(1, keepdim=True)[1]
13              correct += pred.eq(target.view_as(pred)).sum().item()
14      test_loss /= len(test_loader.dataset)
15      test_accuracy = correct / len(test_loader.dataset)
16      return test_loss, test_accuracy
17
18  def eval_adv_test(model, device, test_loader):
19      model.eval()
20      test_loss = 0
21      correct = 0
22      with torch.no_grad():
23          for data, target in test_loader:
24              data, target = data.to(device), target.to(device)
25              data = data.view(data.size(0),28*28)
```

```
26          adv_data = adv_attack(model, data, target, device=
      device)
27              output = model(adv_data)
28              test_loss += F.nll_loss(output, target, size_average=
      False).item()
29              pred = output.max(1, keepdim=True)[1]
30              correct += pred.eq(target.view_as(pred)).sum().item()
31      test_loss /= len(test_loader.dataset)
32      test_accuracy = correct / len(test_loader.dataset)
33      return test_loss, test_accuracy
```

D.2.8   Adversarial Training

Below is the second place needing your implementation, for O2. In the template
code, there is a baseline method. You can replace relevant part of the code as
indicated in the code.

```
1  #train function, you can use adversarial training
2  def train(args, model, device, train_loader, optimizer, epoch):
3      model.train()
4      for batch_idx, (data, target) in enumerate(train_loader):
5          data, target = data.to(device), target.to(device)
6          data = data.view(data.size(0),28*28)
7
8          #use adverserial data to train the defense model
9          #adv_data = adv_attack(model, data, target, device=device
      )
10
11          #clear gradients
12          optimizer.zero_grad()
13
14          #compute loss
15          #loss = F.nll_loss(model(adv_data), target)
16          loss = F.nll_loss(model(data), target)
17
18          #get gradients and update
19          loss.backward()
20          optimizer.step()
21
22  #main function, train the dataset and print train loss, test loss
          for each epoch
23  def train_model():
24      model = Net().to(device)
25
26      #
      ###############################################################
27      ## Note: below is the place you need to edit to implement
      your own training algorithm
28      ##        You can also edit the functions such as train(...).
```

```
29    #
      ###############################################################
30
31    optimizer = optim.SGD(model.parameters(), lr=args.lr)
32    for epoch in range(1, args.epochs + 1):
33        start_time = time.time()
34
35        #training
36        train(args, model, device, train_loader, optimizer, epoch
      )
37
38        #get trnloss and testloss
39        trnloss, trnacc = eval_test(model, device, train_loader)
40        advloss, advacc = eval_adv_test(model, device,
      train_loader)
41
42        #print trnloss and testloss
43        print('Epoch '+str(epoch)+': '+str(int(time.time()-
      start_time))+'s', end=', ')
44        print('trn_loss: {:.4f}, trn_acc: {:.2f}%'.format(trnloss
      , 100. * trnacc), end=', ')
45        print('adv_loss: {:.4f}, adv_acc: {:.2f}%'.format(advloss
      , 100. * advacc))
46
47    adv_tstloss, adv_tstacc = eval_adv_test(model, device,
      test_loader)
48    print('Your estimated attack ability, by applying your attack
       method on your own trained model, is: {:.4f}'.format(1/
      adv_tstacc))
49    print('Your estimated defence ability, by evaluating your own
       defence model over your attack, is: {:.4f}'.format(
      adv_tstacc))
50    ###############################################
51    ## end of training method
52    ###############################################
53
54    #save the model
55    torch.save(model.state_dict(), str(id_)+'.pt')
56    return model
```

D.2.9   Define Distance Metrics

In this competition, we take the L_∞ as the distance measure. You are not allowed to change the code.

```
1  #compute perturbation distance
2  def p_distance(model, train_loader, device):
3      p = []
4      for batch_idx, (data, target) in enumerate(train_loader):
5          data, target = data.to(device), target.to(device)
```

```
6        data = data.view(data.size(0),28*28)
7        data_ = copy.deepcopy(data.data)
8        adv_data = adv_attack(model, data, target, device=device)
9        p.append(torch.norm(data_-adv_data, float('inf')))
10   print('epsilon p: ',max(p))
```

D.2.10   Supplementary Code for Test Purpose

In addition to the above code, we also provide two lines of code for testing purpose. You must comment them out in your submission. The first line is to call the **train_model()** method to train a new model, and the second is to check the quality of attack based on a model.

```
1   #Comment out the following command when you do not want to re-
        train the model
2   #In that case, it will load a pre-trained model you saved in
        train_model()
3   model = train_model()
4
5   #Call adv_attack() method on a pre-trained model
6   #The robustness of the model is evaluated against the infinite-
        norm distance measure
7   #!!! Important: MAKE SURE the infinite-norm distance (epsilon p)
        less than 0.11 !!!
8   p_distance(model, train_loader, device)
```

D.3   Implementation Actions

Below, we summarise the actions that need to be taken for the completion of a submission:

1. You must assign the variable **id_** with your student ID i;
2. You need to update the **adv_attack** function with your adversarial attack method;
3. You may change the hyper-parameters defined in **parser** if needed;
4. You must make sure the perturbation distance less than **0.11**, (which can be computed by **p_distance** function);
5. You need to update the **train_model** function (and some other functions that it called such as **train**) with your own training method;
6. You need to use the line "model = train_model()" to train a model and check whether there is a file i.**pt**, which stores the weights of your trained model;
7. You must submit i.**zip**, which includes two files i.**pt** (saved model) and competition_i.**py** (your script).

D.3.1 Sanity Check

Please make sure that the following constraints are satisfied. Your submission won't be marked if they are not followed.

- Submission file: please follow the naming convention as suggested above.
- Make sure your code can run smoothly.
- Comment out the two lines "model = train_model()" and "p_distance(model, train_loader , device)", which are for test purpose.

D.4 Evaluation Criteria

Assume that, among the submissions S, we have n submissions that can run smoothly and correctly. We can get model M_i by reading the file i.pt. Then, we collect the following matrix

$$
\textbf{Score} = \begin{matrix} \textbf{i} = \textbf{1} \\ \textbf{i} = \textbf{2} \\ \cdots \\ \textbf{i} = \textbf{n} - \textbf{1} \\ \textbf{i} = \textbf{n} \end{matrix} \begin{pmatrix} s_{11} & s_{12} & \cdots & s_{1(n-1)} & s_{1n} \\ s_{21} & s_{22} & \cdots & s_{2(n-1)} & s_{2n} \\ & & & & \\ s_{(n-1)1} & s_{(n-1)2} & \cdots & s_{(n-1)(n-1)} & s_{(n-1)n} \\ s_{n1} & s_{n2} & \cdots & s_{n(n-1)} & s_{nn} \end{pmatrix}
$$
$$
\textbf{j} = \textbf{1} \quad \textbf{j} = \textbf{2} \quad \cdots \quad \textbf{j} = \textbf{n} - \textbf{1} \quad \textbf{j} = \textbf{n}
$$

(D.1)

for the mutual evaluation scores of using M_i to evaluate Atk_j (defined in function **adv_attack**). The score s_{ij} is the **test accuracy** obtained by using **adv_attack** function from the file competition $\_j$.**py** to attack the model from i.pt. From Eq. (D.1), we get j's attacking ability by letting

$$
Attack Ability_j = \sum_{i=1}^{n} \textbf{Score}_{i,j} \tag{D.2}
$$

to be the total of the scores of j-th column. Let **AttackAbility** be the vector of $Attack Ability_j$. Moreover, we get i's defence ability by letting

$$
Defence Ability_i = \sum_{j=1}^{n} \textbf{Score}_{i,j}, \tag{D.3}
$$

Let **DefenceAbility** be the vector of $DefenceAbility_i$. Then, for the vectors **AttackAbility** and **DefenceAbility**, we apply Softmax function to normalise to get

$$\sigma(\frac{1}{\textbf{AttackAbility}}), \sigma(\textbf{DefenceAbility}) \qquad (D.4)$$

Then, the final score for the submission i is

$$FinalScore_i = \sigma(\sigma(\frac{1}{\textbf{AttackAbility}}) + \sigma(\textbf{DefenceAbility}))_i \qquad (D.5)$$

Note that, to reduce the impact of randomness, we may conduct 3 rounds of the above process to get the average $FinalScore_i$ for every submission.

D.5   Q&A

Will the Running Time, Memory, and CPU Usage be Considered in Marking for the Training and Attack?

We only evaluate against a trained model, so no consideration will be given on the running time, memory, and CPU usage in training. For the attack, it will run on marker's computer, so we expect it to run in a reasonable time, e.g., within 1 min for a single image.

Can We Use External Pip Packages for Purposes Like Hyperparameter Tuning?

You can use external libraries as long as they are consistent with the libraries we suggested for this module, and can be installed easily through either pip or annaconda.

You can also use package for e.g., attack, but you have to make sure that the package will run well in normal circumstance (so that the markers can run the package).

Would it Be OK to Search for Popular Algorithm and Then Try to Fix that to My Submission?

Yes, you can. Actually, you are free to take—and adapt—any existing algorithms/implementations—this is also a skill that is nowadays quite useful in ML field. This will probably be the case for most people.

How Would You Make Sure the Competition Is Fair? I am Worrying About that Some Submissions May Try to Overfit the Test Dataset

The test dataset is not open, so nobody has access to the test dataset before competition.

Architectural Search Was Mentioned in Today's Q&A Session, But I Think it may be Unhelpful Since Our Target Network Architecture Is Fixed. The Only Thing I Could Come up with Is to Use Dropouts to Deactivate Some Nodes to Achieve a Pseudo Different Net Structure? Architecture search might not be the most suitable technique to consider here—I mentioned it simply because there was a question I was asked.

Automated Hyperparameter Tunning Was Also Mentioned in Today's Q&A. Also, You Mentioned Several Sessions Ago that We Should Make Sure to Use Some Widely Used Reliable Libraries. Given that I am Using Ray Tune Currently, Do You have any Recommendations on This?

It might be OK to try, if you want to exercise on hyper-parameter tuning. I do not have personal preference on the libraries, and you can select whichever you feel OK with. Just to remind you: for the submission, we only look at the trained weights, so will not be able to consider how you tune the parameters to get the weights.

How Will the Score of Attack/Defense be Measured? Do We Use Loss or Accuracy or Loss & Accuracy to Evaluate the Result?

Please refer to the $p\_distance$ function of the template code for a good understanding of this question.

How Could You Deal with the Connection Problem when Downloading Dataset Through Python IDLE?

Please use CMD (windows) or Terminal (mac) to download the dataset.

Glossary

f	Machine Learning Model
J	Objective Function
θ	Machine Learning Model Parameters
$\mathbf{W}$	Machine Learning Model Weight Matrix
D_{train}	Training Dataset
D_{test}	Test Dataset
$\mathcal{D}$	Input Domain
C	A set of labels
$\mathbf{x}$	A data instance
X_i	The i-th feature of data instance
y	A scalar label of the data instance $\mathbf{x}$
$\hat{y}$	The predictive label of the data instance $\mathbf{x}$ by a classifier
$\mathcal{L}$	Loss function
h	Ground-truth function
PCTL	Probabilistic Computation Tree Logic
MDP	Markov Decision Process
DDPG	Deep Deterministic Policy Gradient
PID	Proportional–Integral–Derivative

© The Author(s), under exclusive license to Springer Nature Singapore Pte Ltd. 2023   307
X. Huang et al., *Machine Learning Safety*, Artificial Intelligence: Foundations,
Theory, and Algorithms, https://doi.org/10.1007/978-981-19-6814-3

References

1. Zeyuan Allen-Zhu, Yuanzhi Li, and Yingyu Liang. Learning and generalization in overparameterized neural networks, going beyond two layers. In Hanna M. Wallach, Hugo Larochelle, Alina Beygelzimer, Florence d'Alché-Buc, Emily B. Fox, and Roman Garnett, editors, *Advances in Neural Information Processing Systems 32: Annual Conference on Neural Information Processing Systems 2019, NeurIPS 2019, December 8-14, 2019, Vancouver, BC, Canada*, pages 6155–6166, 2019.
2. Pierre Alquier, James Ridgway, and Nicolas Chopin. On the properties of variational approximations of gibbs posteriors. *J. Mach. Learn. Res.*, 17:239:1–239:41, 2016.
3. Amany Alshareef, Nicolas Berthier, Sven Schewe, and Xiaowei Huang. Quantifying the importance of latent features in neural networks. In Gabriel Pedroza, José Hernández-Orallo, Xin Cynthia Chen, Xiaowei Huang, Huáscar Espinoza, Mauricio Castillo-Effen, John McDermid, Richard Mallah, and Seán Ó hÉigeartaigh, editors, *Proceedings of the Workshop on Artificial Intelligence Safety 2022 (SafeAI 2022) co-located with the Thirty-Sixth AAAI Conference on Artificial Intelligence (AAAI2022), Virtual, February, 2022*, volume 3087 of *CEUR Workshop Proceedings*. CEUR-WS.org, 2022.
4. Dario Amodei, Chris Olah, Jacob Steinhardt, Paul F. Christiano, John Schulman, and Dan Mané. Concrete problems in AI safety. *CoRR*, abs/1606.06565, 2016.
5. Karol Arndt, Murtaza Hazara, Ali Ghadirzadeh, and Ville Kyrki. Meta reinforcement learning for sim-to-real domain adaptation. In *2020 IEEE International Conference on Robotics and Automation (ICRA)*, pages 2725–2731, 2020.
6. Anish Athalye, Nicholas Carlini, and David A. Wagner. Obfuscated gradients give a false sense of security: Circumventing defenses to adversarial examples. In Jennifer G. Dy and Andreas Krause, editors, *Proceedings of the 35th International Conference on Machine Learning, ICML 2018, Stockholmsmässan, Stockholm, Sweden, July 10-15, 2018*, volume 80 of *Proceedings of Machine Learning Research*, pages 274–283. PMLR, 2018.
7. Yang Bai, Yan Feng, Yisen Wang, Tao Dai, Shu-Tao Xia, and Yong Jiang. Hilbert-based generative defense for adversarial examples. In *Proceedings of the IEEE/CVF International Conference on Computer Vision*, pages 4784–4793, 2019.
8. Raef Bassily, Adam D. Smith, and Abhradeep Thakurta. Private empirical risk minimization: Efficient algorithms and tight error bounds. In *55th IEEE Annual Symposium on Foundations of Computer Science, FOCS 2014, Philadelphia, PA, USA, October 18-21, 2014*, pages 464–473. IEEE Computer Society, 2014.
9. Roberto Battiti. First-and second-order methods for learning: between steepest descent and newton's method. *Neural computation*, 4(2):141–166, 1992.

10. Nicolas Berthier, Amany Alshareef, James Sharp, Sven Schewe, and Xiaowei Huang. Abstraction and symbolic execution of deep neural networks with bayesian approximation of hidden features, 2021.

11. Battista Biggio, Igino Corona, Davide Maiorca, Blaine Nelson, Nedim Šrndić, Pavel Laskov, Giorgio Giacinto, and Fabio Roli. Evasion attacks against machine learning at test time. In *Joint European conference on machine learning and knowledge discovery in databases*, pages 387–402. Springer, 2013.

12. Christopher M Bishop. *Pattern recognition and machine learning.* springer, 2006.

13. Charles Blundell, Julien Cornebise, Koray Kavukcuoglu, and Daan Wierstra. Weight uncertainty in neural network. In *International Conference on Machine Learning*, pages 1613–1622. PMLR, 2015.

14. Aleksandar Botev, Hippolyt Ritter, and David Barber. Practical gauss-newton optimisation for deep learning. In *International Conference on Machine Learning*, pages 557–565. PMLR, 2017.

15. Eric Breck, Neoklis Polyzotis, Sudip Roy, Steven Whang, and Martin Zinkevich. Data validation for machine learning. In Ameet Talwalkar, Virginia Smith, and Matei Zaharia, editors, *Proceedings of Machine Learning and Systems 2019, MLSys 2019, Stanford, CA, USA, March 31 - April 2, 2019*. mlsys.org, 2019.

16. Rudy Bunel, Ilker Turkaslan, Philip HS Torr, Pushmeet Kohli, and M Pawan Kumar. Piecewise linear neural network verification: A comparative study. *arXiv preprint arXiv:1711.00455*, 2017.

17. Nicholas Carlini and David Wagner. Adversarial examples are not easily detected: Bypassing ten detection methods. In *Proceedings of the 10th ACM Workshop on Artificial Intelligence and Security*, pages 3–14. ACM, 2017.

18. Nicholas Carlini and David Wagner. Magnet and" efficient defenses against adversarial attacks" are not robust to adversarial examples. *arXiv preprint arXiv:1711.08478*, 2017.

19. Nicholas Carlini and David Wagner. Towards evaluating the robustness of neural networks. In *Security and Privacy (SP), IEEE Symposium on*, pages 39–57, 2017.

20. Stephanie C. Y. Chan, Sam Fishman, John F. Canny, Anoop Korattikara, and Sergio Guadar-rama. Measuring the reliability of reinforcement learning algorithms. *CoRR*, abs/1912.05663, 2019.

21. Niladri S Chatterji, Behnam Neyshabur, and Hanie Sedghi. The intriguing role of module criticality in the generalization of deep networks. *ICLR2020*, 2020.

22. Tianlong Chen, Zhenyu Zhang, Sijia Liu, Shiyu Chang, and Zhangyang Wang. Robust overfitting may be mitigated by properly learned smoothening. In *International Conference on Learning Representations*, 2020.

23. Chih-Hong Cheng, Georg Nührenberg, and Harald Ruess. Maximum resilience of artificial neural networks. In Deepak D'Souza and K. Narayan Kumar, editors, *Automated Technology for Verification and Analysis*, pages 251–268. Springer, 2017.

24. Paul Christiano, Zain Shah, Igor Mordatch, Jonas Schneider, Trevor Blackwell, Joshua Tobin, Pieter Abbeel, and Wojciech Zaremba. Transfer from simulation to real world through learning deep inverse dynamics model. *arXiv preprint arXiv:1610.03518*, 2016.

25. Edmund M Clarke Jr, Orna Grumberg, Daniel Kroening, Doron Peled, and Helmut Veith. *Model checking*. The MIT Press, 2018.

26. Jiequan Cui, Shu Liu, Liwei Wang, and Jiaya Jia. Learnable boundary guided adversarial training. In *Proceedings of the IEEE/CVF International Conference on Computer Vision*, pages 15721–15730, 2021.

27. Christian Dehnert, Sebastian Junges, Joost-Pieter Katoen, and Matthias Volk. A Storm is coming: A modern probabilistic model checker. In Rupak Majumdar and Viktor Kunĉak, editors, *Computer Aided Verification*, volume 10427 of *LNCS*, pages 592–600, Cham, 2017. Springer.

28. Ambra Demontis, Marco Melis, Maura Pintor, Matthew Jagielski, Battista Biggio, Alina Oprea, Cristina Nita-Rotaru, and Fabio Roli. Why do adversarial attacks transfer? explaining transferability of evasion and poisoning attacks. In *28th USENIX Security Symposium (USENIX Security 19)*, pages 321–338, Santa Clara, CA, August 2019. USENIX Association.

29. Terrance DeVries and Graham W Taylor. Improved regularization of convolutional neural networks with cutout. *arXiv preprint arXiv:1708.04552*, 2017.

30. Gavin Weiguang Ding, Yash Sharma, Kry Yik Chau Lui, and Ruitong Huang. MMA training: Direct input space margin maximization through adversarial training. In *8th International Conference on Learning Representations, ICLR 2020, Addis Ababa, Ethiopia, April 26-30, 2020*. OpenReview.net, 2020.

31. Yukun Ding, Jinglan Liu, Jinjun Xiong, and Yiyu Shi. Revisiting the evaluation of uncertainty estimation and its application to explore model complexity-uncertainty trade-off. In *2020 IEEE/CVF Conference on Computer Vision and Pattern Recognition, CVPR Workshops 2020, Seattle, WA, USA, June 14-19, 2020*, pages 22–31. Computer Vision Foundation / IEEE, 2020.

32. Yi Dong, Xingyu Zhao, and Xiaowei Huang. Dependability analysis of deep reinforcement learning based robotics and autonomous systems. *CoRR*, abs/2109.06523, 2021.

33. Yinpeng Dong, Zhijie Deng, Tianyu Pang, Jun Zhu, and Hang Su. Adversarial distributional training for robust deep learning. In Hugo Larochelle, Marc'Aurelio Ranzato, Raia Hadsell, Maria-Florina Balcan, and Hsuan-Tien Lin, editors, *Advances in Neural Information Processing Systems 33: Annual Conference on Neural Information Processing Systems 2020, NeurIPS 2020, December 6-12, 2020, virtual*, 2020.

34. Yinpeng Dong, Zhijie Deng, Tianyu Pang, Jun Zhu, and Hang Su. Adversarial distributional training for robust deep learning. In H. Larochelle, M. Ranzato, R. Hadsell, M. F. Balcan, and H. Lin, editors, *Advances in Neural Information Processing Systems*, volume 33, pages 8270–8283. Curran Associates, Inc., 2020.

35. Yinpeng Dong, Qi-An Fu, Xiao Yang, Tianyu Pang, Hang Su, Zihao Xiao, and Jun Zhu. Benchmarking adversarial robustness on image classification. In *Proceedings of the IEEE/CVF Conference on Computer Vision and Pattern Recognition*, pages 321–331, 2020.

36. Alexey Dosovitskiy, Germán Ros, Felipe Codevilla, Antonio M. López, and Vladlen Koltun. CARLA: an open urban driving simulator. In *1st Annual Conference on Robot Learning, CoRL 2017, Mountain View, California, USA, November 13-15, 2017, Proceedings*, volume 78 of *Proceedings of Machine Learning Research*, pages 1–16. PMLR, 2017.

37. Simon S. Du, Jason D. Lee, Haochuan Li, Liwei Wang, and Xiyu Zhai. Gradient descent finds global minima of deep neural networks. In Kamalika Chaudhuri and Ruslan Salakhutdinov, editors, *Proceedings of the 36th International Conference on Machine Learning, ICML 2019, 9-15 June 2019, Long Beach, California, USA*, volume 97 of *Proceedings of Machine Learning Research*, pages 1675–1685. PMLR, 2019.

38. Dheeru Dua and Casey Graff. UCI machine learning repository, 2017.

39. Souradeep Dutta, Susmit Jha, Sriram Sanakaranarayanan, and Ashish Tiwari. Output range analysis for deep neural networks. *arXiv preprint arXiv:1709.09130*, 2018.

40. Cynthia Dwork, Vitaly Feldman, Moritz Hardt, Toniann Pitassi, Omer Reingold, and Aaron Roth. Generalization in adaptive data analysis and holdout reuse. In Corinna Cortes, Neil D. Lawrence, Daniel D. Lee, Masashi Sugiyama, and Roman Garnett, editors, *Advances in Neural Information Processing Systems 28: Annual Conference on Neural Information Processing Systems 2015, December 7-12, 2015, Montreal, Quebec, Canada*, pages 2350–2358, 2015.

41. Cynthia Dwork, Vitaly Feldman, Moritz Hardt, Toniann Pitassi, Omer Reingold, and Aaron Leon Roth. Preserving statistical validity in adaptive data analysis. In *Proceedings of the forty-seventh annual ACM symposium on Theory of computing*, pages 117–126, 2015.

42. Gintare Karolina Dziugaite and Daniel M Roy. Computing nonvacuous generalization bounds for deep (stochastic) neural networks with many more parameters than training data. *Conference on Uncertainty in Artificial Intelligence (UAI)*, 2017.

43. Gintare Karolina Dziugaite and Daniel M. Roy. Data-dependent pac-bayes priors via differential privacy. In Samy Bengio, Hanna M. Wallach, Hugo Larochelle, Kristen Grauman, Nicolò Cesa-Bianchi, and Roman Garnett, editors, *Advances in Neural Information Processing Systems 31: Annual Conference on Neural Information Processing Systems 2018, NeurIPS 2018, December 3-8, 2018, Montréal, Canada*, pages 8440–8450, 2018.

44. Ruediger Ehlers. Formal verification of piece-wise linear feed-forward neural networks. In *International Symposium on Automated Technology for Verification and Analysis*, pages 269–286. Springer, 2017.

45. Logan Engstrom, Andrew Ilyas, and Anish Athalye. Evaluating and understanding the robustness of adversarial logit pairing. *arXiv preprint arXiv:1807.10272*, 2018.

46. Logan Engstrom, Dimitris Tsipras, Ludwig Schmidt, and Aleksander Madry. A rotation and a translation suffice: Fooling cnns with simple transformations. *arXiv preprint arXiv:1712.02779*, 2017.

47. Ilenia Epifani, Carlo Ghezzi, Raffaela Mirandola, and Giordano Tamburrelli. Model evolution by run-time parameter adaptation. In *Proc. of the 31st Int. Conf. on Software Engineering*, ICSE '09, pages 111–121, Washington, DC, USA, 2009. IEEE Computer Society.

48. Keinosuke Fukunaga. *Introduction to statistical pattern recognition*. Elsevier, 2013.

49. Fadri Furrer, Michael Burri, Markus Achtelik, and Roland Siegwart. *RotorS—A Modular Gazebo MAV Simulator Framework*, pages 595–625. Springer International Publishing, Cham, 2016.

50. Yarin Gal and Zoubin Ghahramani. Dropout as a bayesian approximation: Representing model uncertainty in deep learning. In *Proceedings of the 33rd International Conference on International Conference on Machine Learning - Volume 48*, ICML'16, page 1050–1059. JMLR.org, 2016.

51. Timon Gehr, Matthew Mirman, Dana Drachsler-Cohen, Petar Tsankov, Swarat Chaudhuri, and Martin Vechev. AI2: Safety and robustness certification of neural networks with abstract interpretation. In *Security and Privacy (SP), 2018 IEEE Symposium on*, 2018.

52. Timon Gehr, Matthew Mirman, Dana Drachsler-Cohen, Petar Tsankov, Swarat Chaudhuri, and Martin Vechev. Ai2: Safety and robustness certification of neural networks with abstract interpretation. In *2018 IEEE Symposium on Security and Privacy (SP)*, pages 3–18. IEEE, 2018.

53. Victor Gergel, Vladimir Grishagin, and Alexander Gergel. Adaptive nested optimization scheme for multidimensional global search. *Journal of Global Optimization*, 66(1):35–51, 2016.

54. Pascal Germain, Alexandre Lacasse, François Laviolette, and Mario Marchand. Pac-bayesian learning of linear classifiers. In *Proceedings of the 26th Annual International Conference on Machine Learning*, pages 353–360, 2009.

55. Xueluan Gong, Yanjiao Chen, Wenbin Yang, Guanghao Mei, and Qian Wang. Inversenet: Augmenting model extraction attacks with training data inversion. In Zhi-Hua Zhou, editor, *Proceedings of the Thirtieth International Joint Conference on Artificial Intelligence, IJCAI 2021, Virtual Event / Montreal, Canada, 19-27 August 2021*, pages 2439–2447. ijcai.org, 2021.

56. Ian J. Goodfellow, Jonathon Shlens, and Christian Szegedy. Explaining and harnessing adversarial examples. In *International Conference on Learning Representations*, 2015.

57. Vladimir Grishagin, Ruslan Israfilov, and Yaroslav Sergeyev. Convergence conditions and numerical comparison of global optimization methods based on dimensionality reduction schemes. *Applied Mathematics and Computation*, 318:270–280, 2018.

58. Jamie Hayes and George Danezis. Learning universal adversarial perturbations with generative models. In *2018 IEEE Security and Privacy Workshops (SPW)*, pages 43–49. IEEE, 2018.

59. Kelly Hayhurst, Dan Veerhusen, John Chilenski, and Leanna Rierson. A practical tutorial on modified condition/decision coverage. Technical report, NASA, 2001.

60. Kaiming He, Xiangyu Zhang, Shaoqing Ren, and Jian Sun. Deep residual learning for image recognition. In *Proceedings of the IEEE conference on computer vision and pattern recognition*, pages 770–778, 2016.

61. Dan Hendrycks and Kevin Gimpel. A baseline for detecting misclassified and out-of-distribution examples in neural networks. In *5th International Conference on Learning Representations, ICLR 2017, Toulon, France, April 24-26, 2017, Conference Track Proceedings*. OpenReview.net, 2017.

62. Dan Hendrycks and Kevin Gimpel. Early methods for detecting adversarial images. In *5th International Conference on Learning Representations, ICLR 2017, Toulon, France, April 24-26, 2017, Workshop Track Proceedings*. OpenReview.net, 2017.

63. Thomas A. Henzinger, Anna Lukina, and Christian Schilling. Outside the box: Abstraction-based monitoring of neural networks. In Giuseppe De Giacomo, Alejandro Catalá, Bistra Dilkina, Michela Milano, Senén Barro, Alberto Bugarín, and Jérôme Lang, editors, *ECAI 2020 - 24th European Conference on Artificial Intelligence, 29 August-8 September 2020, Santiago de Compostela, Spain, August 29 - September 8, 2020 - Including 10th Conference on Prestigious Applications of Artificial Intelligence (PAIS 2020)*, volume 325 of *Frontiers in Artificial Intelligence and Applications*, pages 2433–2440. IOS Press, 2020.

64. Martin Heusel, Hubert Ramsauer, Thomas Unterthiner, Bernhard Nessler, and Sepp Hochreiter. Gans trained by a two time-scale update rule converge to a local nash equilibrium. In *Proceedings of the 31st International Conference on Neural Information Processing Systems*, NIPS'17, page 6629–6640, Red Hook, NY, USA, 2017. Curran Associates Inc.

65. W. Ronny Huang, Jonas Geiping, Liam Fowl, Gavin Taylor, and Tom Goldstein. Metapoison: Practical general-purpose clean-label data poisoning. In H. Larochelle, M. Ranzato, R. Hadsell, M. F. Balcan, and H. Lin, editors, *Advances in Neural Information Processing Systems*, volume 33, pages 12080–12091. Curran Associates, Inc., 2020.

66. Wei Huang, Youcheng Sun, Xingyu Zhao, James Sharp, Wenjie Ruan, Jie Meng, and Xiaowei Huang. Coverage-guided testing for recurrent neural networks. *IEEE Transactions on Reliability*, pages 1–16, 2021.

67. Wei Huang, Xingyu Zhao, and Xiaowei Huang. Embedding and extraction of knowledge in tree ensemble classifiers. *Machine Learning*, 2021.

68. Xiaowei Huang, Daniel Kroening, Wenjie Ruan, James Sharp, Youcheng Sun, Emese Thamo, Min Wu, and Xinping Yi. A survey of safety and trustworthiness of deep neural networks: Verification, testing, adversarial attack and defence, and interpretability. *Computer Science Review*, 37:100270, 2020.

69. Xiaowei Huang, Marta Kwiatkowska, and Maciej Olejnik. Reasoning about cognitive trust in stochastic multiagent systems. *ACM Trans. Comput. Logic*, 20(4), jul 2019.

70. Xiaowei Huang, Marta Kwiatkowska, Sen Wang, and Min Wu. Safety verification of deep neural networks. In *International Conference on Computer Aided Verification*, pages 3–29. Springer, 2017.

71. Laurent Hyafil and Ronald L. Rivest. Constructing optimal binary decision trees is np-complete. *Information Processing Letters*, 5(1):15–17, 1976.

72. Sheikh Rabiul Islam, William Eberle, and Sheikh K. Ghafoor. Towards quantification of explainability in explainable artificial intelligence methods. *CoRR*, abs/1911.10104, 2019.

73. Yangqing Jia, Evan Shelhamer, Jeff Donahue, Sergey Karayev, Jonathan Long, Ross Girshick, Sergio Guadarrama, and Trevor Darrell. Caffe: Convolutional architecture for fast feature embedding. *arXiv preprint arXiv:1408.5093*, 2014.

74. Yue Jia and Mark Harman. An analysis and survey of the development of mutation testing. *IEEE Transactions on Software Engineering*, 37(5):649–678, 2011.

75. Yiding Jiang, Behnam Neyshabur, Hossein Mobahi, Dilip Krishnan, and Samy Bengio. Fantastic generalization measures and where to find them. *arXiv preprint arXiv:1912.02178*, 2019.

76. Gaojie Jin, Xinping Yi, Wei Huang, Sven Schewe, and Xiaowei Huang. Enhancing adversarial training with second-order statistics of weights. *arXiv preprint arXiv:2203.06020*, 2022.

77. Gaojie Jin, Xinping Yi, Liang Zhang, Lijun Zhang, Sven Schewe, and Xiaowei Huang. How does weight correlation affect the generalisation ability of deep neural networks. In *NeurIPS'20*, 2020.

78. Justin Johnson, Alexandre Alahi, and Li Fei-Fei. Perceptual losses for real-time style transfer and super-resolution. In *European Conference on Computer Vision*, pages 694–711. Springer, 2016.

79. Harini Kannan, Alexey Kurakin, and Ian Goodfellow. Adversarial logit pairing. *arXiv preprint arXiv:1803.06373*, 2018.

80. Alex Kantchelian, J. D. Tygar, and Anthony D. Joseph. Evasion and hardening of tree ensemble classifiers. In *Proceedings of the 33nd International Conference on Machine Learning*, volume 48, pages 2387–2396, 2016.

81. Manuel Kaspar, Juan D. Muñoz Osorio, and Juergen Bock. Sim2real transfer for reinforcement learning without dynamics randomization. In *2020 IEEE/RSJ International Conference on Intelligent Robots and Systems (IROS)*, pages 4383–4388, 2020.

82. Guy Katz, Clark Barrett, David L Dill, Kyle Julian, and Mykel J Kochenderfer. Reluplex: An efficient SMT solver for verifying deep neural networks. In *International Conference on Computer Aided Verification*, pages 97–117. Springer, 2017.

83. Nitish Shirish Keskar, Dheevatsa Mudigere, Jorge Nocedal, Mikhail Smelyanskiy, and Ping Tak Peter Tang. On large-batch training for deep learning: Generalization gap and sharp minima. In *5th International Conference on Learning Representations, ICLR 2017, Toulon, France, April 24-26, 2017, Conference Track Proceedings*. OpenReview.net, 2017.

84. Diederik P. Kingma and Jimmy Ba. Adam: A method for stochastic optimization. In *International Conference on Learning Representations*, 2015.

85. K. Kristinsson and G.A. Dumont. System identification and control using genetic algorithms. *IEEE Transactions on Systems, Man, and Cybernetics*, 22(5):1033–1046, 1992.

86. S. Kullback and R.A. Leibler. On information and sufficiency. *Annals of Mathematical Statistics*, 22(1):79–86, 1951.

87. Alexey Kurakin, Ian Goodfellow, Samy Bengio, et al. Adversarial examples in the physical world, 2016.

88. Alexey Kurakin, Ian J. Goodfellow, and Samy Bengio. Adversarial examples in the physical world. *CoRR*, abs/1607.02533, 2016.

89. Marta Kwiatkowska, Gethin Norman, and David Parker. PRISM 4.0: Verification of probabilistic real-time systems. In Ganesh Gopalakrishnan and Shaz Qadeer, editors, *Computer Aided Verification*, volume 6806 of *LNCS*, pages 585–591, Berlin, Heidelberg, 2011. Springer Berlin Heidelberg.

90. Marta Kwiatkowska, Gethin Norman, and David Parker. Probabilistic Model Checking: Advances and Applications. In Rolf Drechsler, editor, *Formal System Verification: State-of-the-Art and Future Trends*, pages 73–121. Springer, Cham, 2018.

91. Balaji Lakshminarayanan, Alexander Pritzel, and Charles Blundell. Simple and scalable predictive uncertainty estimation using deep ensembles. In Isabelle Guyon, Ulrike von Luxburg, Samy Bengio, Hanna M. Wallach, Rob Fergus, S. V. N. Vishwanathan, and Roman Garnett, editors, *Advances in Neural Information Processing Systems 30: Annual Conference on Neural Information Processing Systems 2017, December 4-9, 2017, Long Beach, CA, USA*, pages 6402–6413, 2017.

92. John Langford and Rich Caruana. (not) bounding the true error. *Advances in Neural Information Processing Systems*, 2:809–816, 2002.

93. John Langford and John Shawe-Taylor. Pac-bayes & margins. *Advances in neural information processing systems*, pages 439–446, 2003.

94. Y. Lecun, L. Bottou, Y. Bengio, and P. Haffner. Gradient-based learning applied to document recognition. *Proceedings of the IEEE*, 86(11):2278–2324, 1998.

95. Saehyung Lee, Hyungyu Lee, and Sungroh Yoon. Adversarial vertex mixup: Toward better adversarially robust generalization. In *Proceedings of the IEEE/CVF Conference on Computer Vision and Pattern Recognition*, pages 272–281, 2020.

96. Melanie Lefkowitz. Professor's perceptron paved the way for ai – 60 years too soon, 2019.

97. Gaël Letarte, Pascal Germain, Benjamin Guedj, and François Laviolette. Dichotomize and generalize: Pac-bayesian binary activated deep neural networks. In Hanna M. Wallach, Hugo Larochelle, Alina Beygelzimer, Florence d'Alché-Buc, Emily B. Fox, and Roman Garnett, editors, *Advances in Neural Information Processing Systems 32: Annual Conference on Neural Information Processing Systems 2019, NeurIPS 2019, December 8-14, 2019, Vancouver, BC, Canada*, pages 6869–6879, 2019.

98. Jianlin Li, Jiangchao Liu, Pengfei Yang, Liqian Chen, Xiaowei Huang, and Lijun Zhang. Analyzing deep neural networks with symbolic propagation: Towards higher precision and faster verification. In *SAS2019*, 2019.

99. Renjue Li, Jianlin Li, Cheng-Chao Huang, Pengfei Yang, Xiaowei Huang, Lijun Zhang, Bai Xue, and Holger Hermanns. Prodeep: A platform for robustness verification of deep neural networks. In *Proceedings of the 28th ACM Joint Meeting on European Software Engineering Conference and Symposium on the Foundations of Software Engineering*, ESEC/FSE 2020, page 1630–1634, New York, NY, USA, 2020. Association for Computing Machinery.

100. Timothy P Lillicrap, Jonathan J Hunt, Alexander Pritzel, Nicolas Heess, Tom Erez, Yuval Tassa, David Silver, and Daan Wierstra. Continuous control with deep reinforcement learning. In *ICLR'16*, 2016.

101. B. Littlewood and J. Rushby. Reasoning about the reliability of diverse two-channel systems in which one channel is "possibly perfect". *IEEE Transactions on Software Engineering*, 38(5):1178–1194, 2012.

102. Alessio Lomuscio and Lalit Maganti. An approach to reachability analysis for feed-forward ReLU neural networks. *arXiv preprint arXiv:1706.07351*, 2017.

103. Xingjun Ma, Bo Li, Yisen Wang, Sarah M. Erfani, Sudanthi N. R. Wijewickrema, Grant Schoenebeck, Dawn Song, Michael E. Houle, and James Bailey. Characterizing adversarial subspaces using local intrinsic dimensionality. In *6th International Conference on Learning Representations, ICLR 2018, Vancouver, BC, Canada, April 30 - May 3, 2018, Conference Track Proceedings*. OpenReview.net, 2018.

104. Aleksander Madry, Aleksandar Makelov, Ludwig Schmidt, Dimitris Tsipras, and Adrian Vladu. Towards deep learning models resistant to adversarial attacks. In *6th International Conference on Learning Representations, ICLR 2018, Vancouver, BC, Canada, April 30 - May 3, 2018, Conference Track Proceedings*. OpenReview.net, 2018.

105. Aravindh Mahendran and Andrea Vedaldi. Understanding deep image representations by inverting them. In *IEEE Conference on Computer Vision and Pattern Recognition, CVPR 2015, Boston, MA, USA, June 7-12, 2015*, pages 5188–5196. IEEE Computer Society, 2015.

106. Chengzhi Mao, Ziyuan Zhong, Junfeng Yang, Carl Vondrick, and Baishakhi Ray. Metric learning for adversarial robustness. In Hanna M. Wallach, Hugo Larochelle, Alina Beygelzimer, Florence d'Alché-Buc, Emily B. Fox, and Roman Garnett, editors, *Advances in Neural Information Processing Systems 32: Annual Conference on Neural Information Processing Systems 2019, NeurIPS 2019, December 8-14, 2019, Vancouver, BC, Canada*, pages 478–489, 2019.

107. James Martens and Roger Grosse. Optimizing neural networks with kronecker-factored approximate curvature. In *International conference on machine learning*, pages 2408–2417. PMLR, 2015.

108. Andreas Maurer. A note on the pac bayesian theorem. *arXiv preprint cs/0411099*, 2004.

109. David A McAllester. PAC-bayesian model averaging. In *Proceedings of the twelfth annual conference on Computational learning theory*, pages 164–170, 1999.

110. Dongyu Meng and Hao Chen. Magnet: a two-pronged defense against adversarial examples. In *Proceedings of the 2017 ACM SIGSAC Conference on Computer and Communications Security*, pages 135–147. ACM, 2017.

111. Jan Hendrik Metzen, Tim Genewein, Volker Fischer, and Bastian Bischoff. On detecting adversarial perturbations. In *5th International Conference on Learning Representations, ICLR 2017, Toulon, France, April 24-26, 2017, Conference Track Proceedings*. OpenReview.net, 2017.

112. Volodymyr Mnih, Koray Kavukcuoglu, David Silver, Alex Graves, Ioannis Antonoglou, Daan Wierstra, and Martin Riedmiller. Playing atari with deep reinforcement learning. *arXiv preprint arXiv:1312.5602*, 2013.

113. Seyed-Mohsen Moosavi-Dezfooli, Alhussein Fawzi, Omar Fawzi, and Pascal Frossard. Universal adversarial perturbations. *2017 IEEE Conference on Computer Vision and Pattern Recognition (CVPR)*, pages 86–94, 2017.

114. Seyed-Mohsen Moosavi-Dezfooli, Alhussein Fawzi, and Pascal Frossard. Deepfool: a simple and accurate method to fool deep neural networks. In *Proceedings of the IEEE Conference on Computer Vision and Pattern Recognition*, pages 2574–2582, 2016.

115. Rafael Müller, Simon Kornblith, and Geoffrey E. Hinton. When does label smoothing help? In Hanna M. Wallach, Hugo Larochelle, Alina Beygelzimer, Florence d'Alché-Buc, Emily B. Fox, and Roman Garnett, editors, *Advances in Neural Information Processing Systems 32: Annual Conference on Neural Information Processing Systems 2019, NeurIPS 2019, December 8-14, 2019, Vancouver, BC, Canada*, pages 4696–4705, 2019.

116. Fabio Muratore, Christian Eilers, Michael Gienger, and Jan Peters. Bayesian domain randomization for sim-to-real transfer. *CoRR*, abs/2003.02471, 2020.

117. Taesik Na, Jong Hwan Ko, and Saibal Mukhopadhyay. Cascade adversarial machine learning regularized with a unified embedding. In *International Conference on Learning Representations (ICLR)*, 2018.

118. Vinod Nair and Geoffrey E Hinton. Rectified linear units improve restricted Boltzmann machines. In *Proceedings of the 27th International Conference on Machine Learning (ICML)*, pages 807–814, 2010.

119. Arvind Narayanan and Vitaly Shmatikov. Robust de-anonymization of large sparse datasets. In *2008 IEEE Symposium on Security and Privacy (sp 2008)*, pages 111–125, 2008.

120. Nina Narodytska, Shiva Prasad Kasiviswanathan, Leonid Ryzhyk, Mooly Sagiv, and Toby Walsh. Verifying properties of binarized deep neural networks. In *Proceedings of the Thirty-Second AAAI Conference on Artificial Intelligence*, 2018.

121. Augustus Odena and Ian Goodfellow. TensorFuzz: Debugging neural networks with coverage-guided fuzzing. *arXiv preprint arXiv:1807.10875*, 2018.

122. Tribhuvanesh Orekondy, Bernt Schiele, and Mario Fritz. Knockoff nets: Stealing functionality of black-box models. In *IEEE Conference on Computer Vision and Pattern Recognition, CVPR 2019, Long Beach, CA, USA, June 16-20, 2019*, pages 4954–4963. Computer Vision Foundation / IEEE, 2019.

123. Soham Pal, Yash Gupta, Aditya Shukla, Aditya Kanade, Shirish K. Shevade, and Vinod Ganapathy. Activethief: Model extraction using active learning and unannotated public data. In *The Thirty-Fourth AAAI Conference on Artificial Intelligence, AAAI 2020, The Thirty-Second Innovative Applications of Artificial Intelligence Conference, IAAI 2020, The Tenth AAAI Symposium on Educational Advances in Artificial Intelligence, EAAI 2020, New York, NY, USA, February 7-12, 2020*, pages 865–872. AAAI Press, 2020.

124. Tianyu Pang, Xiao Yang, Yinpeng Dong, Hang Su, and Jun Zhu. Bag of tricks for adversarial training. In *9th International Conference on Learning Representations, ICLR 2021, Virtual Event, Austria, May 3-7, 2021*. OpenReview.net, 2021.

125. Nicolas Papernot, Martín Abadi, Úlfar Erlingsson, Ian J. Goodfellow, and Kunal Talwar. Semi-supervised knowledge transfer for deep learning from private training data. In *5th International Conference on Learning Representations, ICLR 2017, Toulon, France, April 24-26, 2017, Conference Track Proceedings*. OpenReview.net, 2017.

126. Nicolas Papernot, Fartash Faghri, Nicholas Carlini, Ian Goodfellow, Reuben Feinman, Alexey Kurakin, Cihang Xie, Yash Sharma, Tom Brown, Aurko Roy, Alexander Matyasko, Vahid Behzadan, Karen Hambardzumyan, Zhishuai Zhang, Yi-Lin Juang, Zhi Li, Ryan Sheatsley, Abhibhav Garg, Jonathan Uesato, Willi Gierke, Yinpeng Dong, David Berthelot, Paul Hendricks, Jonas Rauber, and Rujun Long. Technical report on the cleverhans v2.1.0 adversarial examples library. *arXiv preprint arXiv:1610.00768*, 2018.

127. Nicolas Papernot, Patrick McDaniel, Xi Wu, Somesh Jha, and Ananthram Swami. Distillation as a defense to adversarial perturbations against deep neural networks. In *Security and Privacy (SP), 2016 IEEE Symposium on*, pages 582–597. IEEE, 2016.

128. Nicolas Papernot, Patrick D. McDaniel, Ian J. Goodfellow, Somesh Jha, Z. Berkay Celik, and Ananthram Swami. Practical black-box attacks against deep learning systems using adversarial examples. *CoRR*, abs/1602.02697, 2016.

129. Nicolas Papernot, Patrick D. McDaniel, Somesh Jha, Matt Fredrikson, Z. Berkay Celik, and Ananthram Swami. The limitations of deep learning in adversarial settings. In *Security and Privacy (EuroS&P), 2016 IEEE European Symposium on*, pages 372–387. IEEE, 2016.

130. Emilio Parrado-Hernández, Amiran Ambroladze, John Shawe-Taylor, and Shiliang Sun. Pac-bayes bounds with data dependent priors. *The Journal of Machine Learning Research*, 13(1):3507–3531, 2012.

131. Kexin Pei, Yinzhi Cao, Junfeng Yang, and Suman Jana. DeepXplore: Automated whitebox testing of deep learning systems. In *Proceedings of the 26th Symposium on Operating Systems Principles*, pages 1–18. ACM, 2017.

132. Kexin Pei, Yinzhi Cao, Junfeng Yang, and Suman Jana. Towards practical verification of machine learning: The case of computer vision systems. *arXiv preprint arXiv:1712.01785*, 2017.

133. María Pérez-Ortiz, Omar Rivasplata, John Shawe-Taylor, and Csaba Szepesvári. Tighter risk certificates for neural networks. *Journal of Machine Learning Research*, 22, 2021.

134. Claudia Perlich, Foster Provost, and Jeffrey S. Simonoff. Tree induction vs. logistic regression: A learning-curve analysis. *J. Mach. Learn. Res.*, 4(null):211–255, December 2003.

135. Roberto Pietrantuono, Peter Popov, and Stefano Russo. Reliability assessment of service-based software under operational profile uncertainty. *Reliability Engineering & System Safety*, 204:107193, 2020.

136. SA Piyavskii. An algorithm for finding the absolute extremum of a function. *USSR Computational Mathematics and Mathematical Physics*, 12(4):57–67, 1972.

137. Omid Poursaeed, Isay Katsman, Bicheng Gao, and Serge J. Belongie. Generative adversarial perturbations. In *Conference on Computer Vision and Pattern Recognition*, 2018.

138. Ramya Ramakrishnan, Ece Kamar, Debadeepta Dey, Eric Horvitz, and Julie Shah. Blind spot detection for safe sim-to-real transfer. *J. Artif. Intell. Res.*, 67:191–234, 2020.

139. Hippolyt Ritter, Aleksandar Botev, and David Barber. A scalable laplace approximation for neural networks. In *6th International Conference on Learning Representations, ICLR 2018-Conference Track Proceedings*, volume 6. International Conference on Representation Learning, 2018.

140. Omar Rivasplata, Vikram M Tankasali, and Csaba Szepesvari. Pac-bayes with backprop. *arXiv preprint arXiv:1908.07380*, 2019.

141. Robotis. Robotis(2019) turtlebot3 – e-manual, waffle pi. [Online] https://emanual.robotis.com/docs/en/platform/turtlebot3/overview/. (Accessed on 02 August 2021).

142. Olaf Ronneberger, Philipp Fischer, and Thomas Brox. U-net: Convolutional networks for biomedical image segmentation. In *International Conference on Medical image computing and computer-assisted intervention*, pages 234–241. Springer, 2015.

143. RTCA. Do-178c, software considerations in airborne systems and equipment certification. 2011.

144. Wenjie Ruan, Xiaowei Huang, and Marta Kwiatkowska. Reachability analysis of deep neural networks with provable guarantees. In *IJCAI*, pages 2651–2659, 2018.

145. Wenjie Ruan, Min Wu, Youcheng Sun, Xiaowei Huang, Daniel Kroening, and Marta Kwiatkowska. Global robustness evaluation of deep neural networks with provable guarantees for the hamming distance. In Sarit Kraus, editor, *Proceedings of the Twenty-Eighth International Joint Conference on Artificial Intelligence, IJCAI 2019, Macao, China, August 10-16, 2019*, pages 5944–5952. ijcai.org, 2019.

146. Wenjie Ruan, Min Wu, Youcheng Sun, Xiaowei Huang, Daniel Kroening, and Marta Kwiatkowska. Global robustness evaluation of deep neural networks with provable guarantees for the hamming distance. In Sarit Kraus, editor, *Proceedings of the Twenty-Eighth International Joint Conference on Artificial Intelligence, IJCAI 2019, Macao, China, August 10-16, 2019*, pages 5944–5952. ijcai.org, 2019.

147. John Rushby. Software verification and system assurance. In *7th Int. Conf. on Software Engineering and Formal Methods*, pages 3–10, Hanoi, Vietnam, 2009. IEEE.

148. Aniruddha Saha, Akshayvarun Subramanya, and Hamed Pirsiavash. Hidden trigger backdoor attacks. In *The Thirty-Fourth AAAI Conference on Artificial Intelligence, AAAI 2020, The Thirty-Second Innovative Applications of Artificial Intelligence Conference, IAAI 2020, The Tenth AAAI Symposium on Educational Advances in Artificial Intelligence, EAAI 2020, New York, NY, USA, February 7-12, 2020*, pages 11957–11965. AAAI Press, 2020.

149. Sebastian Schelter, Dustin Lange, Philipp Schmidt, Meltem Celikel, Felix Biessmann, and Andreas Grafberger. Automating large-scale data quality verification. *Proc. VLDB Endow.*, 11(12):1781–1794, aug 2018.

150. Ali Shafahi, W. Ronny Huang, Mahyar Najibi, Octavian Suciu, Christoph Studer, Tudor Dumitras, and Tom Goldstein. Poison frogs! targeted clean-label poisoning attacks on neural networks. In *Proceedings of the 32nd International Conference on Neural Information Processing Systems*, NIPS'18, page 6106–6116, Red Hook, NY, USA, 2018. Curran Associates Inc.

151. Shital Shah, Debadeepta Dey, Chris Lovett, and Ashish Kapoor. Airsim: High-fidelity visual and physical simulation for autonomous vehicles. In Marco Hutter and Roland Siegwart, editors, *Field and Service Robotics*, pages 621–635, Cham, 2018. Springer International Publishing.

152. Adrian J Shepherd. *Second-order methods for neural networks: Fast and reliable training methods for multi-layer perceptrons*. Springer Science & Business Media, 2012.

153. Karen Simonyan, Andrea Vedaldi, and Andrew Zisserman. Deep inside convolutional networks: Visualising image classification models and saliency maps. In Yoshua Bengio and Yann LeCun, editors, *2nd International Conference on Learning Representations, ICLR 2014, Banff, AB, Canada, April 14-16, 2014, Workshop Track Proceedings*, 2014.

154. Nitish Srivastava, Geoffrey Hinton, Alex Krizhevsky, Ilya Sutskever, and Ruslan Salakhutdinov. Dropout: A simple way to prevent neural networks from overfitting. *Journal of Machine Learning Research*, 15(56):1929–1958, 2014.

155. Matthew Staib and Stefanie Jegelka. Distributionally robust deep learning as a generalization of adversarial training. *NIPS workshop on Machine Learning and Computer Security*, 2017.

156. Lorenzo Strigini and Andrey Povyakalo. Software fault-freeness and reliability predictions. In *SafeComp2013*, pages 106–117, 2013.

157. Ting Su, Ke Wu, Weikai Miao, Geguang Pu, Jifeng He, Yuting Chen, and Zhendong Su. A survey on data-flow testing. *ACM Computing Surveys*, 50(1):5:1–5:35, March 2017.

158. Youcheng Sun, Xiaowei Huang, and Daniel Kroening. Testing deep neural networks. *CoRR*, abs/1803.04792, 2018.

159. Youcheng Sun, Xiaowei Huang, Daniel Kroening, James Shap, Matthew Hill, and Rob Ashmore. Structural test coverage criteria for deep neural networks. 2018.

160. Youcheng Sun, Min Wu, Wenjie Ruan, Xiaowei Huang, Marta Kwiatkowska, and Daniel Kroening. Concolic testing for deep neural networks. In *Automated Software Engineering (ASE), 33rd IEEE/ACM International Conference on*, 2018.

161. Youcheng Sun, Min Wu, Wenjie Ruan, Xiaowei Huang, Marta Kwiatkowska, and Daniel Kroening. Deepconcolic: Testing and debugging deep neural networks. In *41st ACM/IEEE International Conference on Software Engineering (ICSE2019)*, 2018.

162. Richard S Sutton and Andrew G Barto. *Reinforcement learning: An introduction*. MIT press, 2018.

163. Christian Szegedy, Vincent Vanhoucke, Sergey Ioffe, Jon Shlens, and Zbigniew Wojna. Rethinking the inception architecture for computer vision. In *Proceedings of the IEEE conference on computer vision and pattern recognition*, pages 2818–2826, 2016.

164. Christian Szegedy, Wojciech Zaremba, Ilya Sutskever, Joan Bruna, Dumitru Erhan, Ian Goodfellow, and Rob Fergus. Intriguing properties of neural networks. *arXiv preprint arXiv:1312.6199*, 2013.

165. Christian Szegedy, Wojciech Zaremba, Ilya Sutskever, Joan Bruna, Dumitru Erhan, Ian Goodfellow, and Rob Fergus. Intriguing properties of neural networks. In *In ICLR*. Citeseer, 2014.

166. Niklas Thiemann, Christian Igel, Olivier Wintenberger, and Yevgeny Seldin. A strongly quasiconvex pac-bayesian bound. In *International Conference on Algorithmic Learning Theory*, pages 466–492. PMLR, 2017.

167. Aimo Torn and Antanas Zilinskas. *Global Optimization*. Springer-Verlag New York, Inc., New York, NY, USA, 1989.

168. Florian Tramèr, Alexey Kurakin, Nicolas Papernot, Ian Goodfellow, Dan Boneh, and Patrick McDaniel. Ensemble Adversarial Training: Attacks and Defenses. In *International Conference on Learning Representations*, 2018.

169. Florian Tramèr, Fan Zhang, Ari Juels, Michael K. Reiter, and Thomas Ristenpart. Stealing machine learning models via prediction apis. In *Proceedings of the 25th USENIX Conference on Security Symposium*, SEC'16, page 601–618, USA, 2016. USENIX Association.

170. René Traoré, Hugo Caselles-Dupré, Timothée Lesort, Te Sun, Natalia Díaz Rodríguez, and David Filliat. Continual reinforcement learning deployed in real-life using policy distillation and sim2real transfer. *CoRR*, abs/1906.04452, 2019.

171. V. N. Vapnik and A. Ya. Chervonenkis. *On the Uniform Convergence of Relative Frequencies of Events to Their Probabilities*, pages 11–30. Springer International Publishing, Cham, 2015.

172. Fu Wang, Peipei Xu, Xiaowei Huang, and Wenjie Ruan. Georobust: Evaluating geometric robustness of neural networks with provable guarantees. In *submitted*, 2022.

173. Jingkang Wang, Yang Liu, and Bo Li. Reinforcement learning with perturbed rewards. In *The Thirty-Fourth AAAI Conference on Artificial Intelligence, AAAI 2020, The Thirty-Second Innovative Applications of Artificial Intelligence Conference, IAAI 2020, The Tenth AAAI Symposium on Educational Advances in Artificial Intelligence, EAAI 2020, New York, NY, USA, February 7-12, 2020*, pages 6202–6209. AAAI Press, 2020.

174. Yisen Wang, Difan Zou, Jinfeng Yi, James Bailey, Xingjun Ma, and Quanquan Gu. Improving adversarial robustness requires revisiting misclassified examples. In *International Conference on Learning Representations*, 2019.

175. Gellért Weisz, Philip Amortila, and Csaba Szepesvári. Exponential lower bounds for planning in mdps with linearly-realizable optimal action-value functions. In Vitaly Feldman, Katrina Ligett, and Sivan Sabato, editors, *Algorithmic Learning Theory, 16-19 March 2021, Virtual Conference, Worldwide*, volume 132 of *Proceedings of Machine Learning Research*, pages 1237–1264. PMLR, 2021.

176. Max Welling and Yee W Teh. Bayesian learning via stochastic gradient langevin dynamics. In *Proceedings of the 28th international conference on machine learning (ICML-11)*, pages 681–688. Citeseer, 2011.

177. T.-W. Weng, H. Zhang, P.-Y. Chen, J. Yi, D. Su, Y. Gao, C.-J. Hsieh, and L. Daniel. Evaluating the Robustness of Neural Networks: An Extreme Value Theory Approach. In *ICLR2018*, 2018.

178. Matthew Wicker, Xiaowei Huang, and Marta Kwiatkowska. Feature-guided black-box safety testing of deep neural networks. In *International Conference on Tools and Algorithms for the Construction and Analysis of Systems*, pages 408–426. Springer, 2018.

179. Dongxian Wu, Yisen Wang, Shu-Tao Xia, James Bailey, and Xingjun Ma. Skip connections matter: On the transferability of adversarial examples generated with resnets. In *8th International Conference on Learning Representations, ICLR 2020, Addis Ababa, Ethiopia, April 26-30, 2020*. OpenReview.net, 2020.

180. Dongxian Wu, Shu-Tao Xia, and Yisen Wang. Adversarial weight perturbation helps robust generalization. *Advances in Neural Information Processing Systems*, 33, 2020.

181. Min Wu, Matthew Wicker, Wenjie Ruan, Xiaowei Huang, and Marta Kwiatkowska. A game-based approximate verification of deep neural networks with provable guarantees. *Theor. Comput. Sci.*, 807:298–329, 2020.

182. Weiming Xiang, Hoang-Dung Tran, and Taylor T Johnson. Output reachable set estimation and verification for multi-layer neural networks. *IEEE Transactions on Neural Networks and Learning Systems*, 29:5777–5783, 2018.

183. Chaowei Xiao, Jun-Yan Zhu, Bo Li, Warren He, Mingyan Liu, and Dawn Song. Spatially transformed adversarial examples. In *6th International Conference on Learning Representations, ICLR 2018, Vancouver, BC, Canada, April 30 - May 3, 2018, Conference Track Proceedings*. OpenReview.net, 2018.

184. Cihang Xie and Alan Yuille. Intriguing properties of adversarial training at scale. In *International Conference on Learning Representations*, 2019.

185. Weilin Xu, David Evans, and Yanjun Qi. Feature squeezing: Detecting adversarial examples in deep neural networks. In *Network and Distributed System Security Symposium (NDSS)*, 2018.

186. Pengfei Yang, Jianlin Li, Jiangchao Liu, Cheng-Chao Huang, Renjue Li, Liqian Chen, Xiaowei Huang, and Lijun Zhang. Enhancing robustness verification for deep neural networks via symbolic propagation. *Form. Asp. Comput.*, 33(3):407–435, jun 2021.

187. Ziqi Yang, Jiyi Zhang, Ee-Chien Chang, and Zhenkai Liang. Neural network inversion in adversarial setting via background knowledge alignment. In *Proceedings of the 2019 ACM SIGSAC Conference on Computer and Communications Security*, CCS '19, page 225–240, New York, NY, USA, 2019. Association for Computing Machinery.

188. Jason Yosinski, Jeff Clune, Anh Mai Nguyen, Thomas J. Fuchs, and Hod Lipson. Understanding neural networks through deep visualization. *CoRR*, abs/1506.06579, 2015.

189. Honggang Yu, Kaichen Yang, Teng Zhang, Yun-Yun Tsai, Tsung-Yi Ho, and Yier Jin. Cloudleak: Large-scale deep learning models stealing through adversarial examples. In *27th Annual Network and Distributed System Security Symposium, NDSS 2020, San Diego, California, USA, February 23-26, 2020*. The Internet Society, 2020.

190. Yang Yu. Towards sample efficient reinforcement learning. In *IJCAI*, pages 5739–5743, 2018.

191. Matthew D. Zeiler and Rob Fergus. Visualizing and understanding convolutional networks. In *European Conference on Computer Vision (ECCV)*, 2014.

192. Haichao Zhang and Jianyu Wang. Defense against adversarial attacks using feature scattering-based adversarial training. In H. Wallach, H. Larochelle, A. Beygelzimer, F. d'Alché-Buc, E. Fox, and R. Garnett, editors, *Advances in Neural Information Processing Systems*, volume 32. Curran Associates, Inc., 2019.

193. Hongyang Zhang, Yaodong Yu, Jiantao Jiao, Eric Xing, Laurent El Ghaoui, and Michael Jordan. Theoretically principled trade-off between robustness and accuracy. In *International Conference on Machine Learning*, pages 7472–7482. PMLR, 2019.

194. Linjun Zhang, Zhun Deng, Kenji Kawaguchi, Amirata Ghorbani, and James Zou. How does mixup help with robustness and generalization? In *9th International Conference on Learning Representations, ICLR 2021, Virtual Event, Austria, May 3-7, 2021*. OpenReview.net, 2021.

195. Yuheng Zhang, Ruoxi Jia, Hengzhi Pei, Wenxiao Wang, Bo Li, and Dawn Song. The secret revealer: Generative model-inversion attacks against deep neural networks. In *2020 IEEE/CVF Conference on Computer Vision and Pattern Recognition, CVPR 2020, Seattle, WA, USA, June 13-19, 2020*, pages 250–258. Computer Vision Foundation / IEEE, 2020.

196. Wenshuai Zhao, Jorge Peña Queralta, Li Qingqing, and Tomi Westerlund. Ubiquitous distributed deep reinforcement learning at the edge: Analyzing byzantine agents in discrete action spaces. *Procedia Computer Science*, 177:324–329, 2020. The 11th International Conference on Emerging Ubiquitous Systems and Pervasive Networks (EUSPN 2020) / The 10th International Conference on Current and Future Trends of Information and Communication Technologies in Healthcare (ICTH 2020) / Affiliated Workshops.

197. Wenshuai Zhao, Jorge Peña Queralta, Li Qingqing, and Tomi Westerlund. Towards closing the sim-to-real gap in collaborative multi-robot deep reinforcement learning. *CoRR*, abs/2008.07875, 2020.

198. Xingyu Zhao, Alec Banks, James Sharp, Valentin Robu, David Flynn, Michael Fisher, and Xiaowei Huang. A safety framework for critical systems utilising deep neural networks. In *SafeComp2020*, pages 244–259, 2020.

199. Xingyu Zhao, Wei Huang, Alec Banks, Victoria Cox, David Flynn, Sven Schewe, and Xiaowei Huang. Assessing reliability of deep learning through robustness evaluation and operational testing. In *SafeComp2021*, 2021.

200. Xingyu Zhao, Wei Huang, Vibhav Bharti, Yi Dong, Victoria Cox, Alec Banks, Sen Wang, Sven Schewe, and Xiaowei Huang. Reliability assessment and safety arguments for machine learning components in assuring learning-enabled autonomous systems. *CoRR*, abs/2112.00646, 2021.

201. Tianhang Zheng, Changyou Chen, and Kui Ren. Distributionally adversarial attack. *Proceedings of the AAAI Conference on Artificial Intelligence*, 33(01):2253–2260, Jul. 2019.

202. Chen Zhu, W. Ronny Huang, Hengduo Li, Gavin Taylor, Christoph Studer, and Tom Goldstein. Transferable clean-label poisoning attacks on deep neural nets. In Kamalika Chaudhuri and Ruslan Salakhutdinov, editors, *Proceedings of the 36th International Conference on Machine Learning*, volume 97 of *Proceedings of Machine Learning Research*, pages 7614–7623. PMLR, 09–15 Jun 2019.

203. Hong Zhu, Patrick AV Hall, and John HR May. Software unit test coverage and adequacy. *ACM Computing Surveys*, 29(4):366–427, 1997.

204. Barret Zoph and Quoc V. Le. Neural architecture search with reinforcement learning. In *5th International Conference on Learning Representations, ICLR 2017, Toulon, France, April 24-26, 2017, Conference Track Proceedings*. OpenReview.net, 2017.

Printed in the United States
by Baker & Taylor Publisher Services